SPRINGER TRACTS IN MODERN PHYSICS

Ergebnisse
der exakten Natur-
wissenschaften

Volume **74**

Springer-Verlag Berlin Heidelberg GmbH 1974

Manuscripts for publication should be addressed to:

G. HÖHLER, Institut für Theoretische Kernphysik der Universität, 75 Karlsruhe 1, Postfach 6380

Proofs and all correspondence concerning papers in the process of publication should be addressed to:

E. A. NIEKISCH, Institut für Grenzflächenforschung und Vakuumphysik der Kernforschungsanlage Jülich, 517 Jülich, Postfach 365

Library of Congress Cataloging in Publication Data
Main entry under title:

Solid state physics.

(Springer tracts in modern physics; v. 74)
Includes bibliographical references and index.
CONTENTS: Bauer, G.: Determination of electron
temperatures and of hot electron distribution functions
in semiconductors. — Borstel, G., Falge, H. J. and
Otto, A.: Surface and bulk phonon-polaritons observed
by attenuated total reflection.
1. Energy-band theory of solids. 2. Phonons.
3. Polaritons. I. Bauer, Günther, 1942 —
Determination of electron temperatures and of hot
electron distribution functions in semiconductors. 1974.
II. Borstel, G.: Surface and bulk phonon-polaritons
observed by attenuated total reflection. 1974.
III. Series.
QC1. lS797 vol. 74 [QCl76.8.E4] 539ᶜ. 08s [530.4ʹl] 74-23605

ISBN 978-3-662-15873-9   ISBN 978-3-540-37868-6 (eBook)
DOI 10.1007/978-3-540-37868-6

Solid-State Physics

Contents

Determination of Electron Temperatures and of Hot Electron Distribution Functions in Semiconductors
G. BAUER   1

Surface and Bulk Phonon-Polaritons Observed by Attenuated Total Reflection
G. BORSTEL, H. J. FALGE, and A. OTTO   107

Classified Index of Authors and Titles (Volumes 36—74)   149

Determination of Electron Temperatures and of Hot Electron Distribution Functions in Semiconductors

G. BAUER

Contents

1. Introduction   1
2. Determination of Electron Temperatures   4
   2.1. General Considerations   4
      2.1.1. Electron Temperature Model   9
   2.2. Experimental Methods   13
      2.2.1. Dependence of the Mobility on the Electric Field   13
      2.2.2. Hot Electron Hall Effect and Magnetoresistance   18
      2.2.3. Variation of the g-Factor of Conduction Electrons in a Non-Parabolic Band   22
      2.2.4. Hot-Electron Shubnikov de Haas Effect   25
      2.2.5. Time Resolved Observation of Increasing Electron Temperatures . .   31
      2.2.6. Investigations of the Noise Temperature   35
      2.2.7. Thermoelectric Power of Hot Electrons   36
      2.2.8. Burstein-Shift of the Optical Absorption Edge   38
      2.2.9. Hot-Electron Faraday Effect and Birefrigence   42
3. Determination of Hot Electron Distribution Functions   49
   3.1. Theoretical Foundation   49
      3.1.1. Monte Carlo Method   51
      3.1.2. The Iterative Method   53
   3.2. Experimental Methods   55
      3.2.1. Modulation of Intervalence Band Absorption by an Electric Field . .   55
      3.2.2. Determination of the Anisotropy of the Distribution Function by Electric Field Induced Dichroism   63
      3.2.3. Radiative Recombination from Electric Field Excited Hot Carriers .   71
      3.2.4. Optical Interband Absorption in Degenerate Materials   75
      3.2.5. Inelastic Light Scattering from Hot Electrons   78
4. Comparison of Experimental and Numerical Results – Conclusions   83
   4.1. Electron Temperature Model   83
   4.2. Calculated Distribution Functions   94
5. Summary   96
List of Symbols   98
References 101

1. Introduction

Under the influence of an electric field free carriers in semiconductors will gain energy from the field and a steady state will be achieved if the energy gain from the field equals the energy loss to the lattice. This energy loss is provided e.g. by the electron-phonon interaction which transfers energy from the free carriers (electrons or holes) to the lattice. Only in the limit of vanishing electric field the mean carrier energy will stay at its thermal equilibrium value ε_L. At high fields however the mean carrier energy $\langle \varepsilon \rangle$ will be different from ε_L. Often a temperature is associated with $\langle \varepsilon \rangle$, called the "electron temperature" T_e which deviates from the lattice temperature T under high field conditions. As a consequence Ohms law will not be obeyed in high electric fields if the momentum relaxation time τ_m is energy dependent. It has become customary to use the term "hot electron effects" for the description of high field phenomena in semiconductors.

Whereas the theoretical foundations were already developed more than 30 years ago [1], the first experimental work on high field transport in a homogenous semiconductor was published in 1951 [2][1].

Since then a considerable amount of work, both theoretical and experimental, has been devoted to this subject. Primarily, investigations of high field transport yield information on the details of the electron phonon interaction, the interaction of the carriers with impurities and the interaction of the carriers among themselves [3–11].

Beside this, new phenomena have been found: e.g. the anisotropy of the conductivity in cubic crystals, current instabilities due to negative differential resistivity at high fields caused by the population of higher lying conduction band minima (Gunn-effect), modulation of infrared radiation, generation of acoustic flux which is related to a strong disturbance of the phonon distribution, avalanche effects, etc. Some of these phenomena have already led to technological applications (e.g. Gunn elements, avalanche transit time diodes) [8, 10].

Several review articles on high field effects have been published [3–11] and a book entitled: "High Field Transport in Semiconductors" by Conwell [10]. In this book experimental and theoretical investigations which were done until 1967 are presented in detail. In a later review by Asche and Sarbei [8] main emphasis was led on the high field transport properties of the elemental semiconductors Ge and Si. In the meantime much effort was devoted to investigations of high field effects in new materials e.g. III–V compounds which have become available as high purity single crystals. New experimental techniques have been used, extending former ones to low-temperatures, high pressures, high magnetic fields and a wide range of carrier concentrations. Fast pulse techniques allow nowadays a time resolution of the order of 1 ns. Also

---

[1] Essentially the basic theory has been developed even earlier since the scattering of electrons by acoustic phonons is similar to the scattering of electrons by heavy ions in a plasma. See: Chapman, S., Cowling, T. G.: Mathematical theory of non-uniform gases. Cambridge: University Press, 1939.

new theoretical techniques have been developed which permit numerical solutions of the Boltzmann transport equation, based either on a Monte Carlo method or on iterative methods [11].

High field transport properties are calculated by a solution of the Boltzmann transport equation which was originally established as a statistical theory of dilute gases. Electrons and phonons may be considered as the gaslike particles in a solid. Although there are certain limitations in the application of the Boltzmann equation to transport problems it is almost the only method for a description of a system far from equilibrium as recently stated by Kubo [12].

Due to the complicated nature of the Boltzmann transport equation, being an integro-differential equation, it is usually not possible to solve it analytically. Then several assumptions have to be made. The most extensively used assumption concerns the form of the distribution function which is assumed to be a Maxwell-Boltzmann distribution at a temperature T_e different from the lattice temperature T and centered at the drift velocity v_d.

If we restrict the investigations to homogeneous semiconductors up to fields where the carrier concentration does not change, experimental evidence for hot electron effects is got from $j(E)$ characteristics. Beside these experiments, which yield the change of the drift velocity of the carriers in high fields, sometimes a determination of the mean carrier energy and the high field distribution function is possible.

Since the determination of $j(E)$ characteristics has already been widely discussed [3–10], we will restrict this paper to the description of experimental methods which directly yield information on the dependence of T_e or $\langle \varepsilon \rangle$ on the electric field and on methods for determining distribution functions.

In Chapter 2 we give first a brief account on the theory on which the determination of the electron temperature is based and then describe the various experimental methods. In Chapter 3 the same is presented for the hot electron distribution functions. In Chapter 4 a comparison of experimental data on various materials investigated under quite different ranges of electric fields, concentrations and temperatures is made and the results are collected and discussed.

It is attempted to present also the experimental setups as far as they have been published in order to facilitate the comparison of the inherent possibilities and limitations of the different experimental methods. The main data are compiled in figures and in tabular form together with the experimental conditions so that a quick orientation on the whole scope of this review is made possible. Direct solutions of the Boltzmann equation for certain scattering mechanisms are described in Conwell's book [10] and in Ref. [7, 8] where also the detailed formalism for approximate solutions of the Boltzmann equation using the drifted Maxwell-Boltzmann distribution can be found. Numerical methods, both the Monte Carlo calculation applied to high field transport and iterative methods have been reviewed by Fawcett [11]. Thus we shall describe only the basic principles of the different methods and compare the results with the experiments.

2. Determination of Electron Temperatures

2.1. General Considerations

An experimental investigation of transport properties usually yields information on the mean velocity and mean energy of the whole ensemble of electrons in the semiconductors. In order to deduce theoretically these mean values the distribution function must be known. The kinetic equation for the distribution function is the Boltzmann transport equation under certain conditions [12].

The distribution function $f(k, r, t)$ is defined in such a way that $(2/8\pi^3) f(k, r, t) dk dr$ is the number of electrons in the element of volume $dk dr$ in phase space at time t. We will mainly deal with spatially uniform conditions and therefore the distribution function will not depend on r. The electron density n is given by

$$n = (2/8\pi^3) \int f(k, t) dk . \tag{1}$$

The Boltzmann transport equation is formally written

$$(\partial f(k, t)/\partial t) = (\partial f(k, t)/\partial t)_{\text{field}} + (\partial f(k, t)/\partial t)_{\text{collision}} . \tag{2}$$

In the steady state the left side equals zero and the change of the distribution due to the external fields is balanced by the change due to collisions.

In thermal equilibrium at zero field $f(k)$ will be the Fermi distribution

$$f_0(k) = \{\exp[(\varepsilon(k) - \varepsilon_F)/k_B T] + 1\}^{-1} \tag{3}$$

ε being the electron energy and ε_F the Fermi energy.

For weakly doped semiconductors and at high temperatures $(\varepsilon_F < 0)$ f_0 can be approximated by the Maxwell-Boltzmann distribution (MB)

$$f_0(\varepsilon) = (n/N_c) e^{-\varepsilon/k_B T} \tag{4}$$

where N_c is a density of states factor given by

$$N_c = 2(2\pi m k_B T/h^2)^{3/2}$$

for a parabolic isotropic band with the effective mass m and $\varepsilon = \hbar^2 k^2/2m$.

If only an electric field is applied, the field term of the Boltzmann equation (BE) is given by

$$(\partial f/\partial t)_{\text{field}} = -(e\,\boldsymbol{E}/\hbar)\,\nabla_{\!k} f \,. \tag{5}$$

The collision term will be a sum of various contributions

$$(\partial f/\partial t)_{\text{collision}} = (\partial f/\partial t)_{\text{electron-lattice}} + (\partial f/\partial t)_{\text{electron-impurity}} \\ + (\partial f/\partial t)_{\text{electron-electron}} \,. \tag{6}$$

It follows from the BE that the distribution function in the case where an electric field is applied will differ from the thermal distribution function. This deviation will depend on the nature of the various scattering processes which may be nearly elastic or strongly inelastic and cause preferential small or large angle scattering. Of course it will also depend on the magnitude of the applied electric field.

Generally the collision term is given by the number of electrons which are scattered out of an element of volume $d^3 k$ around k and the number of electrons scattered from all other volume elements $d^3 k'$ into k

$$(\partial f/\partial t)_{\text{c}} = (V/8\pi^3) \int_{\text{zone}} [W(\boldsymbol{k}',\boldsymbol{k})\,f(\boldsymbol{k}')\,(1-f(\boldsymbol{k})) \\ - W(\boldsymbol{k},\boldsymbol{k}')\,f(\boldsymbol{k})\,(1-f(\boldsymbol{k}'))]\,d^3 k'$$

where $W(\boldsymbol{k}',\boldsymbol{k})$ is the probability per time that an electron makes a transition from the state $\boldsymbol{k}'$ to the state $\boldsymbol{k}$ and has the dimension of a reciprocal time. The factors $(1-f(\boldsymbol{k}))$ and $(1-f(\boldsymbol{k}'))$ account for the number of vacant states.

For non-degenerate conditions above equation simplifies to[2]

$$(\partial f(\boldsymbol{k})/\partial t)_{\text{c}} = -f(\boldsymbol{k}) \int W(\boldsymbol{k},\boldsymbol{k}')\,d^3 k' + \int f(\boldsymbol{k}')\,W(\boldsymbol{k}',\boldsymbol{k})\,d^3 k' \,. \tag{7}$$

Thus inserting Eq. (7) in Eq. (2) the BE becomes an integro-differential equation. The complication in solving this equation arises from the fact that the second term of the collision integral depends on the distribution function $f(\boldsymbol{k}')$. In order to calculate how many electrons are scattered into a volume element around $\boldsymbol{k}$, the distribution function $f(\boldsymbol{k}')$ must already be known.

To illustrate the collision term we consider phonon scattering: 4 processes will change the occupation probability of a state with wave vector $\boldsymbol{k}$. If time dependent perturbation theory is used, the collision

---

[2] We use as abbreviations f for field and c for collision.

term is given by [10]

$$(\partial f(k)/\partial t)_c = (2\pi/\hbar) \sum_q [|(k, N_q + 1 | H' | k + q, N_q)|^2$$

$$\cdot \delta(\varepsilon_k - \varepsilon_{k+q} + \hbar\omega_q) f(k+q)$$

$$+ |(k, N_q - 1 | H' | k - q, N_q)|^2 \delta(\varepsilon_k - \varepsilon_{k-q} - \hbar\omega_q) f(k-q) \quad (8)$$

$$- |(k - q, N_q + 1 | H' | k, N_q)|^2 \delta(\varepsilon_{k-q} - \varepsilon_k + \hbar\omega_q) f(k)$$

$$- |(k + q, N_q - 1 | H' | k, N_q)|^2 \delta(\varepsilon_{k+q} - \varepsilon_k - \hbar\omega_q) f(k)]$$

where the first two terms correspond to a scattering into the state k by emission from $(k + q)$ or absorption from $(k - q)$ of $\hbar\omega_q$ and the third and fourth term correspond to the scattering out of the state k by emission into $(k - q)$ and by absorption into $(k + q)$. q represents the wave vector of the phonon, $\hbar\omega_q$ the phonon energy. The δ-functions conserve energy and the transitions are proportional to the square of the appropriate matrix elements in which H' represents the perturbing potential. N_q is the steady state number of phonons which is in thermal equilibrium given by the Bose-Einstein statistics at the temperature T

$$\bar{N}_q = \{\exp(\hbar\omega_q/k_B T) - 1\}^{-1} . \tag{9}$$

Thus, generally the collision term of the BE will have a complicated structure and the resulting integro-differential equation cannot be solved exactly. For low electric field strengths, the disturbance of the distribution from the thermal equilibrium will be small and thus $f(k)$ may be represented by

$$f(k) = f_0(k) + f_1(k) = f_0(\varepsilon) + f_1(\varepsilon)\cos\theta \tag{10}$$

where $f_1 \ll f_0$ and where f_0 is approximated by the Maxwell Boltzmann (MB) distribution at the lattice temperature T. $f_0(k)$ is symmetric around $k = 0$ and thus only $f_1(k)$ will be responsible for the current. In Eq. (10) θ denotes the angle between k and E. Under Ohmic conditions, the BE is solved using Eq. (10) either by making additional assumptions concerning the collision term or by using a variational procedure [4, 7].

In high electric fields, attempts for finding a solution of the BE can also be based on an expansion of $f(k)$ similar to Eq. (10). However, f_0 may not more be large in comparison to f_1 and additionally f_0 may deviate from the MB distribution. Thus additional terms in the expansion of $f(k)$ are needed. Often the distribution function is expanded in a series of Legendre polynomials $P_n(\cos\theta)$ [10, 11]

$$f(k) = \sum_{n=0}^{\infty} f_n(\varepsilon) P_n(\cos\theta) \tag{11}$$

where $P_0(\cos\theta) = 1$, $P_1(\cos\theta) = \cos\theta$, $P_2(\cos\theta) = (3\cos^2\theta - 1)/2$, and where the following normalization is used [13, 14][3]

$$\int_{-1}^{+1} P_n^2(\cos\theta)\,d(\cos\theta) = 2(2n+1)^{-1}.$$

If the expansion Eq. (11) is inserted into the field and collision term of the BE the following expressions are obtained [4]

$$(\partial f/\partial t)_f = (eE/\hbar) \sum_n \{[nk^{n-1}/(2n-1)]\,(d/dk)\,(k^{1-n}f_{n-1})$$

$$+ [(n+1)/(2n+3)\,k^{n+2}]\,(d/dk)\,(k^{n+2}f_{n+1})\}\,P_n(\cos\theta)$$

and

$$(\partial f/\partial t)_c = -(V/8\pi^3) \sum_n \int d^3k'\,\{f_n(k)\,W(k, k') - f_n(k')\,W(k', k)$$

$$\cdot P_n[(\cos(k, k')]\}\,P_n(\cos\theta).$$

Inserting these equations in Eq. (2) a system of equations is obtained by summing terms of equal $P_n(\cos\theta)$ dependence. Of course, further approximations are necessary if this problem shall be solved analytically [4, 7, 8, 10].

An approach for finding a solution of the BE can be made in the following way:

(i) The distribution function in the presence of a high electric field is assumed to be of MB type with a temperature T_e as parameter which is different from the lattice temperature T. The distribution is centered at a certain k_d which corresponds to the drift velocity v_d. It has to be assumed that carrier-carrier scattering is dominant and momentum and energy gained from the field are randomized.

(ii) The distribution function in the presence of high fields is adequately represented by $f_0 + f_1\cos\theta$ (diffusion approximation). f_0 and f_1 are both determined from the BE. It is thus assumed that the high field distribution function is only weakly anisotropic in k-space. If in addition certain limitations on the collision term are made, analytical solutions can be found for special cases: e.g. acoustic phonon scattering in the equipartition approximation [5, 8, 10]. In the equipartition approximation Eq. (9) for N_q is replaced for $\hbar\omega_q \ll k_B T$ by

$$\bar{N}_q + \tfrac{1}{2} = k_B T/\hbar\omega_q. \tag{12}$$

(iii) No assumptions concerning the shape of the distribution function nor the collision term are made and the BE is solved using the iterative method.

---

[3] Sometimes also a surface spherical harmonics expansion is used [13, 14].

(iv) The path of a single electron in k-space is computed by a Monte Carlo technique and the distribution function is derived. The Methods (iii) and (iv) are numerical methods and are discussed in Section 3.1.1 and 3.1.2[4].

In order to understand qualitatively when the Methods (i) or (ii) can be applied we briefly discuss the dominant scattering mechanisms: acoustic phonon scattering (both via deformation potential coupling and via piezoelectric coupling), non-polar and polar optical phonon scattering, intervalley scattering (scattering between different conduction or valence band valleys), and ionized impurity scattering. If the interactions are essentially elastic and if the electrons are not scattered predominantly in a small angle, then the energy gained from the field will cause mainly an increase of the mean "thermal" velocity. The drift velocity will remain low compared with the thermal velocity and therefore the diffusion approximation will be adequate $(f_0 \gg f_1)$. Acoustic phonon scattering which involves phonons near $k = 0$ is approximately an elastic process for semiconductors like Ge, Si, GaAs, InSb etc. if the lattice temperature T exceeds 20 K. This follows from energy and momentum conservation [10]. Ionized impurity scattering is also elastic due to the large mass ratio of impurities and electrons, but due to the Coulomb interaction the momentum change of carriers of small energy will be larger compared with carriers of higher energy. Interactions with phonons at the edge of the Brillouin zone, and with optical phonons will cause changes in the electron energy corresponding to 50 K–700 K. Thus it can be expected that at energies higher than the particular phonon energy a decrease of the occupation probability, a "depopulation", in the carrier distribution function will occur if the lattice temperature or the mean carrier energy is not too high. Then the scattering process with high energetic acoustic or optical phonons is again quasielastic and therefore a diffusion approximation will be adequate. If however the scattering is also preferentially in a small angle and not randomizing, the drift velocity may become comparable to the thermal velocity and the diffusion approximation will be a poor one. This is the case for polar optical phonon scattering where in k-space even a narrowing of the distribution function perpendicular to the field direction will occur. Also intervalley scattering between valleys of different energy (nonequivalent intervalley scattering) may disturb the spherical shape of the distribution function considerably.

---

[4] If only deviations from Ohm's law proportional E^2 are considered ("warm electron range") a variational method first used by Adawi, I.: Phys. Rev. **115**, 1152 (1959); **120**, 118 (1960), has proved to yield results in excellent agreement with experiments in n-Ge. In this calculation the distribution function is set equal to a MB distribution times a series of Legendre polynomials. The coefficients are expanded as a power series in E.

It is necessary to anticipate a change of the phonon distribution too in high field transport since the energy fed into the electron ensemble is dissipated mainly by the emission of phonons. The influence of the disturbed phonon distribution on the electrical conductivity depends on a phonon relaxation time τ_{ph} which describes the relaxation of the disturbed distribution N_q into the thermal Bose-Einstein statistics $\bar{N}_q$ [10]

$$(\partial N_q(t)/\partial t) = - [N_q(t) - \bar{N}_q]/\tau_{ph}(q).$$

Whereas at high lattice temperatures ($T > 77$ K) the phonon relaxation time is so short that even in high field transport N_q can be considered to be the thermal equilibrium distribution function, at low temperatures the disturbance of N_q may have a measurable effect on the transport properties [5].

2.1.1. Electron Temperature Model

In order to justify the electron temperature model carrier-carrier scattering is considered. By this scattering mechanism the electron system as a whole does not change its momentum or energy since no transfer to the lattice occurs. However, this mechanism is responsible for a redistribution of the energy and the momentum gained from the field. Thus it is of major importance for the shape of the distribution function. If carrier-carrier scattering dominates over electron-phonon or electron-impurity interaction it is reasonable to assume that the distribution function will be of MB type. The importance of carrier-carrier scattering will depend on the carrier concentration and for the electron temperature model two ranges are of interest according to Conwell [10]:

(i) carrier concentration which are high enough that the energy exchange between the carriers exceeds that between the carriers and the lattice: f_0 is of MB type, T_e replacing T,

(ii) even higher carrier concentrations for which also the momentum exchange among the carriers exceeds the transfer by all other scattering mechanisms. In this case not only f_0 is determined by carrier-carrier scattering but f, which will be given by

$$f(k) \sim \exp[-\hbar^2(k - k_d)^2/2m k_B T_e]. \tag{13}$$

[5] For a detailed discussion of the influence of a disturbed phonon distribution on carrier heating see: Kocevar, P.: J. Phys. C (Solid State Physics) 5, 3349 (1972) and Perrin, N., Budd, H.: Phys. Rev. B6, 1359 (1972).

It is centered at the drift velocity $v_d = \hbar k_d/m = \mu E$ (μ being the mobility)[6]. For small v_d compared to the average thermal velocity Eq. (13) can be expanded to yield in a diffusion approximation

$$f(k) \sim \exp(-\varepsilon/k_B T_e) \{1 + k E(\hbar \mu/k_B T_e)\}. \tag{14}$$

In the original paper on the "displaced MB distribution function" by Fröhlich and Paranjape [15] a comparison was made for the energy loss rate of a single electron due to electron-electron (e-e) scattering and acoustic phonon scattering. For the e-e scattering, a result of Pines [16] was used derived for the case of a single electron loosing its energy to much lower energetic electrons. According to this treatment e-e scattering should predominate over accoustic phonon scattering for concentrations higher than n_e

$$n_e = (4\pi)^{-1} (\varepsilon^{3/2} m^{3/2} u^2 \varkappa^2/k_B T e^4 \tau_a) \tag{15}$$

where u is the longitudinal sound velocity, $\varkappa$ the dielectric constant and τ_a the mean free time for acoustic phonon scattering. For concentrations higher than n_e an energy randomized distribution should result. Later Hearn [17], and especially Hasegawa and Yamashita [18] and Dykman and Tomchuk [19] used a more elaborate technique, calculating the rate of change of $f(k)$ due to electron-electron collisions with the following relation for $(\partial f/\partial t)_{e\text{-}e}$

$$(\partial f/\partial t)_{e\text{-}e} = \int [f(k_1') f(k_2') - f(k_1) f(k_2)] \, dw \tag{16}$$

where dw is the scattering probability of carriers with wave vector k_1 and k_2 into states k_1' and k_2'. To evaluate dw it was assumed that the electrons interact via a screened Coulomb potential. The resulting complicate expressions can be found in Ref. [8]. Expressions similar to Eq. (15) for optical and polar optical scattering are given in Ref. [20].

In order that the momentum exchange between the carriers dominates that between the carriers and phonons or impurities, higher carrier concentrations are needed than those required for an energy randomized distribution [20]. According to Stratton [20] the carrier concentration which is necessary for a distribution according to Eq. (13) should be higher than $(k_B T/m u^2) n_e$ if acoustic phonons are considered.

We will now briefly describe the solution of the BE for the displaced MB. According to Fröhlich and Paranjape [15] the balance of energy

---

[6] There is a possibility of a decrease of T_e with E ("carrier cooling") due to a narrowing of the distribution function perpendicular to E for predominant optical phonon scattering. Calculations based on the displaced MB distribution [Paranjape, V. V., de Alba, E.: Proc. Phys. Soc. (London) 85, 945 (1960)] led to a decrease of $T_e(E)$. For experiments see Section 2.2.9.

and momentum is used for the determination of the electron temperature and of the drift velocity

$$0 = \int k \{(\partial f/\partial t)_f + (\partial f/\partial t)_c\} \, d^3 k$$
$$0 = \int \varepsilon \{(\partial f/\partial t)_f + (\partial f/\partial t)_c\} \, d^3 k \tag{17}$$

where the abbreviation c for collision and f for field was used. For $E \parallel z$ direction

$$-(\partial f/\partial t)_f = (e E_z/\hbar) (\partial f/\partial k_z)$$

thus

$$e E_z 4\pi^3 n = -\hbar \int k (\partial f/\partial t)_c \, d^3 k \tag{18a}$$

and

$$\int \varepsilon (\partial f/\partial t)_f \, d^3 k = -(e E_z/\hbar) (\hbar^2/2m) \int k^2 (\partial f/\partial k_z) \, d^3 k = e E_z v_d 4\pi^3 n$$
$$e E_z v_d 4\pi^3 n = -\int \varepsilon (\partial f/\partial t)_c \, d^3 k. \tag{18b}$$

The equations are solved by inserting for $(\partial f/\partial t)_c$ the appropriate expressions and by using the displaced MB for f. The calculation is simplified if the expansion Eq. (14) can be used for f. In this diffusion approximation only f_0 enters in the energy balance equation [Eq. (18b)] and only f_1 in the momentum balance Eq. (18a) since only even terms contribute to Eq. (18b) and only odd to Eq. (18a). The two equations are coupled since both have k_d or v_d as a variable. From the first one k_d is calculated and inserted into the energy balance equation which will lead to a relation between T_e and E. The explicit calculation based on the displaced MB for acoustic phonon, non-polar and polar optical phonon scattering is described in Ref. [15, 20]. Ionized impurity scattering is discussed in Ref. [8].

If the carrier concentration is not high enough so that the displaced MB distribution cannot be used but sufficient for an energy randomized distribution and if a relaxation time τ exists for the scattering mechanisms, a further simplification can be made according to Conwell [10]. Then the collision term is given by $-(f - f_0)/\tau$, where $1/\tau(k) = (V/8\pi^3)$ $\cdot \int W(k, k') \, d^3 k'$. By insertion of these expressions into the BE, $(f - f_0)$ $= f_1$ can be expressed as a function of $f_0 (f_1 = -(e\tau(\varepsilon)/\hbar) E (\partial f_0(k)/\partial k))$ and f_0 is assumed to be of MB type depending on T_e. As in the ohmic case, the momentum relaxation time τ_m is calculated and found to be for simple parabolic bands

$$\tau_m(T, T_e) = (4/3\sqrt{\pi}) \int_0^\infty dx \, x^{3/2} \tau(T, x) \exp(-x) \tag{19}$$

where $x = \varepsilon/k_B T_e$. The dependence on the lattice temperature results from $W(k, k')$ which depends on $N_q(T)$. For several scattering mechanisms, the sum over all $W_i(k, k')$ has to be made which is equivalent to the addition of the individual reciprocal τ_i. $T_e(E)$ is calculated from the energy balance equation $e\mu E^2 = P$ where P represents the energy loss rate to the phonons (see Section 4.1). P is evaluated in terms of T_e.

Since we are also concerned with time dependent changes of the mean carrier energy and drift velocity, we derive the balance equations from the BE in which $f = f(k, t)$ for $E \parallel z$. In order to have expressions for macroscopic quantities like drift velocity or mean energy, the BE is multiplied with $\hbar k$ or ε and then integrated over momentum space [21, 22]. The time dependent BE is given by

$$[\partial f(k, t)/\partial t] + (e E_z/\hbar) [\partial f(k, t)/\partial k_z] = (\partial f/\partial t)_c . \tag{20}$$

We use $\langle Q \rangle$ as an abbreviation for

$$\langle Q \rangle = \int f(k, t) Q(k) \, \mathrm{d}^3 k / \int f(k, t) \, \mathrm{d}^3 k . \tag{21}$$

By multiplying the BE with $Q(k)$ and integrating over $\mathrm{d}^3 k$ we obtain

$$\langle (\partial f(k, t)/\partial t) Q(k) \rangle = (\partial/\partial t) (\langle Q(k) \rangle \, 4\pi^3 n)$$

$$\langle (e E_z/\hbar) (\partial f/\partial k_z) Q(k) \rangle = - 4\pi^3 n e \hbar^{-1} E_z \langle \partial Q(k)/\partial k_z \rangle$$

$$\langle (\partial f/\partial t)_c Q(k) \rangle = \int Q(k) (\partial f/\partial t)_c \, \mathrm{d}^3 k .$$

If now for $Q(k) \hbar k = m v$ or ε is inserted we obtain

$$\mathrm{d}(m v_d)/\mathrm{d} t = e E_z + (\hbar/4\pi^3 n) \int k (\partial f/\partial t)_c \, \mathrm{d}^3 k \tag{22}$$

$$\mathrm{d}\langle \varepsilon \rangle/\mathrm{d} t = e E_z v_d + (4\pi^3 n)^{-1} \int \varepsilon (\partial f/\partial t)_c \, \mathrm{d}^3 k . \tag{23}$$

For the momentum balance equation one may define a momentum relaxation time τ_m if the collision term can be replaced by $-(f - f_0)/\tau$.

$$\tau_m = - m v_d [(\hbar/4\pi^3 n) \int k (\partial f/\partial t)_c \, \mathrm{d}^3 k]^{-1} .$$

In the energy balance equation we may define an energy relaxation time

$$\tau_\varepsilon = - (\langle \varepsilon \rangle - \varepsilon_L) [(4\pi^3 n)^{-1} \int \varepsilon (\partial f/\partial t)_c \, \mathrm{d}^3 k]^{-1}$$

where $\langle \varepsilon \rangle$ is defined according to Eq. (21) and ε_L is the thermal energy of a carrier in equilibrium with the lattice. We arrive at the expressions

$$\mathrm{d}(m v_d)/\mathrm{d} t = e E_z - m v_d/\tau_m \tag{24}$$

and

$$\mathrm{d}\langle \varepsilon \rangle/\mathrm{d} t = e E_z v_d - (\langle \varepsilon \rangle - \varepsilon_L)/\tau_\varepsilon . \tag{25}$$

In an analysis of a time dependent hot electron experiment these two equations have to be solved and yield the dependence of the drift velocity and of the mean energy on the electric field.

We still have to establish a connection between the mean carrier energy and the electron temperature. In a diffusion approximation only f_0 has to be considered. If a MB distribution is used and a simple band model considered

$$\langle \varepsilon \rangle = \frac{\int \varepsilon \exp(-\varepsilon/k_B T_e)\, d^3 k}{\int \exp(-\varepsilon/k_B T_e)\, d^3 k} = \frac{\int_0^\infty \varepsilon^{3/2} \exp(-\varepsilon/k_B T_e)\, d\varepsilon}{\int_0^\infty \varepsilon^{1/2} \exp(-\varepsilon/k_B T_e)\, d\varepsilon} \tag{26}$$

$$= \frac{k_B T_e \Gamma(5/2)}{\Gamma(3/2)} = \frac{3}{2} k_B T_e.$$

For degenerate conditions the Fermi distribution function has to be used and

$$\langle \varepsilon \rangle = k_B T_e F_{3/2}(\eta)/F_{1/2}(\eta) \tag{27}$$

where $\eta = \varepsilon_F/k_B T_e$ is the reduced Fermienergy and

$$F_k(\eta) = \int_0^\infty x^k [\exp(x - \eta) + 1]^{-1}\, dx \tag{28}$$

are tabulated functions [23]. For $\eta > 20$, $F_k(\eta)$ can be approximated by $F_k(\eta) = \eta^{k+1}/(k+1)$ thus yielding $\langle \varepsilon \rangle = (3/5)\varepsilon_{F_0}$ the well known result for metals at $T = 0$.

At the end of this chapter, in Table 1, the methods for finding a solution of the high field transport problem are summarized.

2.2. Experimental Methods

2.2.1. Dependence of the Mobility on the Electric Field

A very simple method for determining the field dependence of the electron temperature or the mean carrier energy is based on a comparison of the field dependence of the mobility μ at a fixed low lattice temperature with the dependence of the ohmic mobility μ_0 on the lattice temperature T. This comparison is shown in Fig. 1.

The following assumptions are made in this procedure:

a) The electron distribution function f_0 in the presence of an electric field is adequately be represented by a Maxwellian or Fermi distribution function with a temperature T_e.

Table 1. Solution of high field Boltzmann transport equation

Analytical methods (assumptions concerning f)		Numerical methods (no assumptions concerning f)	
a) Electron temperature model $f(\mathbf{k}) \propto \exp[-\hbar^2(\mathbf{k} - \mathbf{k}_d)/2mk_B T_e]$. Balance equations: $\int \mathbf{k}\{(\partial f/\partial t)_f + (\partial f/\partial t)_c\}\,d^3k = 0$ $\int \varepsilon\{(\partial f/\partial t)_f + (\partial f/\partial t)_c\}\,d^3k = 0$ for f_0 and f_1. $f_0 \propto \exp{-(\varepsilon/k_B T_e)}$ $f_1 = -(e\tau(\varepsilon) E/\hbar)(\partial f_0(\mathbf{k})/\partial \mathbf{k})$. $E(T_e)$ from energy balance $e\mu E^2 = (2/8\pi^3 n)\int f(\mathbf{k})(d\varepsilon/dt)\,d^3k$ $(d\varepsilon/dt) = (V/8\pi^3)[\int \hbar\omega_q W(\mathbf{k}, \mathbf{k}+\mathbf{q})\,d^3q - \int \hbar\omega_q W(\mathbf{k}, \mathbf{k}-\mathbf{q})\,d^3q]$.	b) Diffusion approximation $f(\mathbf{k}) = f_0(\varepsilon) + f_1(\varepsilon)\cos\theta \quad \theta = \sphericalangle\, \mathbf{k}, E$. Solution of BE $(\partial f/\partial t)_f + (\partial f/\partial t)_c = 0$ for f_0 and f_1. Analytically only possible if further approximations concerning the collision term are made e.g. equipartition approximation for phonon occupation number and restrictions on scattering mechanisms.	a) Monte Carlo method path of a single electron in k-space $\mathbf{k} = \mathbf{k}_0 + e E(t - t_0)/\hbar$ terminated according to three probability distribution functions (i) for scattering event to occur, (ii) for particular scattering process, (iii) for choosing $\mathbf{k}'$ after scattering event. $f(\mathbf{k})$ obtained from time which the electron spends in each cell of k-space using the ergodic theorem. Method equivalent to a solution of the BE.	b) Iterative method: Integro-differential BE is reduced to a differential equation by $g_{n+1}(\mathbf{k}) = \int f_n(\mathbf{k}') W(\mathbf{k}', \mathbf{k})\,d^3k'$ then $(eE/\hbar)(\partial f_{n+1}(\mathbf{k})/\partial \mathbf{k})$ $+ f_{n+1}(\mathbf{k}) \lambda(\mathbf{k}) = g_{n+1}(\mathbf{k})$ where $\lambda(\mathbf{k}) = \int W(\mathbf{k}, \mathbf{k}')\,d^3k$ method converges to $f(\mathbf{k})$.

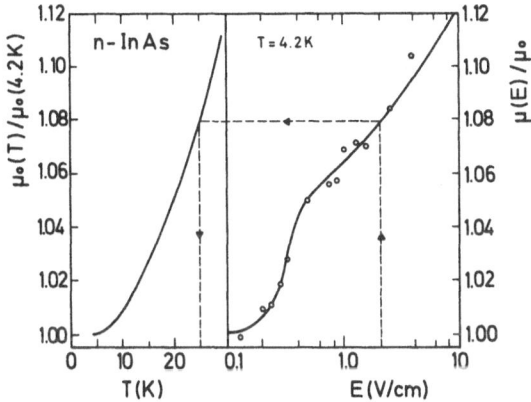

Fig. 1. Determination of the electron temperature from mobility measurements

b) Impurity scattering dominates the momentum relaxation. The impurity relaxation time increases with increasing energy, and has no explicit dependence on lattice temperature. In contrast to this, phonon scattering always depends on the phonon occupation number N_q and will therefore be influenced by the mean carrier energy as well as by the lattice temperature independent of the nature of the particular phonons involved in the scattering process.

Suitable materials for the application of this method are some n-type III–V compounds which have almost spherical conduction band minima at Γ point. Semiconductors with a many valley band-structure or with warped surfaces of constant energy (most of the p-type materials) are unsuitable since the heating of the carriers will then depend on the direction of E with respect to the crystallographic axes [5, 6, 8, 10]. In addition elemental semiconductors like Ge and Si always show a decrease of the carrier concentration (freeze-out) at low temperatures since the activation energies of doping elements in these semiconductors are of the order of 10 meV and higher. Electric fields then cause impact ionization of frozen out electrons in donor sites and the carrier concentration will change.

In III–V compounds ionized impurity scattering limits the mobility at lattice temperatures T approximately smaller than 40 K. n-InSb exhibits above a carrier concentration of 10^{14} cm^{-3} no thermal freeze-out effect [24], so that impact ionization and neutral impurity scattering may be neglected. Several authors have determined the change of T_e with electric field in n-InSb at liquid He temperatures [25–37]. Figure 2 shows results for n-InSb samples with carrier concentrations ranging from 1×10^{14} to 3.2×10^{14} cm^{-3} and for lattice temperatures between

1.35 and 10 K. Independent of the different carrier concentrations and the different degree of compensation as evidenced by different mobilities (Table 2), the data all show a steep increase of T_e for fields between about 0.1 and 0.15 V/cm and a tendency for a smaller increase beyond 0.2 V/cm. In order to explain this rapid rise and the kink at 0.2 V/cm two explanations have been put forward: one is based on the fact that at low mean energies the interaction with acoustic phonons will be responsible for the energy loss, whereas at higher $T_e (>16 \text{ K})$ a more effective energy loss process becomes important [26–37]. On the other hand it was argued that the field dependent conductivity which also shows a kink, is caused by a transfer from electrons in an impurity band like state to the conduction band where they have a higher mobility [25].

Attempts to make Hall measurements in order to get information whether the carrier concentration is constant or not in the range of electric fields applied, have been made by Miyazawa [25], by Manevàl et al. [26] and by Sandercock [32]. Only very small magnetic fields of the order of 100 G have to be applied in order to avoid magnetic freze-out effects. According to the Yafet-Keyes-Adams theory [39] already at 2.03 kG the condition $\hbar\omega_c = 2Ry$ holds where Ry is the Rydberg ionization energy of a hydrogenlike donor and $\hbar\omega_c$ the cyclotron resonance frequency. More refined theories which take into account nonparabolicity, the degree of compensation, and excited impurity states yield a steep increase of the ionization energy of the donors in InSb with magnetic field [40].

Whereas Maneval et al. [26] deduced from their experiments that the carrier density was constant within 1.5% in the range of fields up to 100 mV/cm, Miyazawa [25] found a maximum in $R_B(E)/R_B$ (4.2 K) at about 200 mV/cm approximately 20% higher than R_B, R_B being the Hall coefficient. From the Hall data he concluded that the conduction took not only place in the conduction band. However Sandercock [32] and Crandall [41] still found other explanations for the maximum of $R_B(E)$ as discussed in Section 2.2.2.

The real situation in n-InSb with a carrier concentration of about 10^{14} cm^{-3} will be even more complicated at temperatures lower than about 4.2 K. At these doping levels tails in the density of states of the conduction band are present [42, 43] which originate from statistical doping fluctuations. Overlap of electronic wave functions on neighbouring impurity sites causes broadened impurity levels which form an impurity band and overlap with the density of states tails. If a magnetic field is applied the distribution of the electrons among the conduction band-, tail-, and impurity states will be altered. By using somewhat higher temperatures, the thermal energy of the carriers will be higher and therefore the probability that the relevant conduction process is

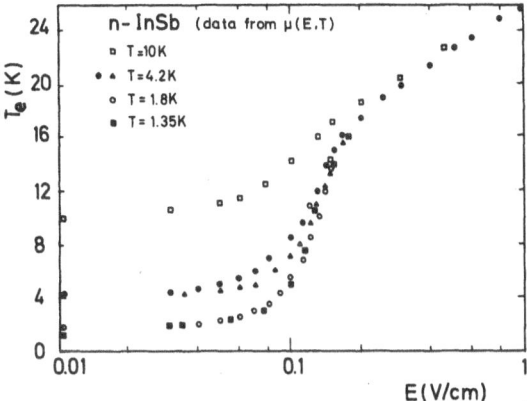

Fig. 2. Dependence of T_e on E in n-InSb. $\square$: $n = 3.2 \times 10^{14}$ cm^{-3}, Ref. [38], $\triangle$, $\bigcirc$: $n = 1 \times 10^{14}$ cm^{-3}, Ref. [25], $\bullet$: $n = 1.19 \times 10^{14}$ cm^{-3}, Ref. [26], $\blacksquare$: $n = 1.25 \times 10^{14}$ cm^{-3}, Ref. [27]

restricted to the conduction band. We have included a curve for n-InSb in Fig. 2 at $T = 10$ K which exhibits no deviation from the $T_e(E)$ curves for fields higher than 200 mV/cm indicating an effective energy loss process for electron temperatures higher than 20 K.

In analysing the data the temperature dependence of the mobility according to the Brooks Herring expression for degenerate statistics is used [44]

$$\mu_{0,I}(\eta) = \frac{2 F_2(\eta)}{F_{1/2}(\eta)} \frac{\sqrt{2} \varkappa^2 (k_B T_e)^{3/2}}{\pi e^3 m^{1/2} N_I g(\overline{x}_{1,\eta})} \tag{29}$$

where $\overline{x}_{s,\eta}$ is determined from

$$[\overline{x}_{s,\eta} - \tfrac{3}{2}(s+1)] = [\overline{x}_{s,\eta} + \tfrac{3}{2}(s+1)] \exp(\eta - \overline{x}_{a,\eta})$$

and

$$g(\overline{x}_{s,\eta}) = \ln(1 + z) - z/(1 + z)$$

where

$$z = \frac{4 \varkappa m (k_B T_e)^2}{\pi e^2 \hbar^2 n} \frac{F_{1/2}(\eta)}{F_{-1/2}(\eta)} x_{s,\eta}.$$

This treatment, based on the Born approximation, is valid for $z \gg 1$.

For the samples of Fig. 2 η is about 0 or higher so that degeneracy has to be considered.

For highly degenerate semiconductors ($\eta \gg 20$) Eq. (29) yields, with the approximation discussed after Eq. (28), only a diminishing de-

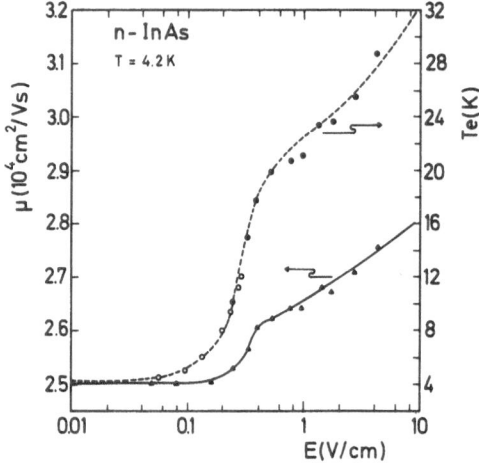

Fig. 3. Dependence of mobility (▲) and T_e (○, ●) on E for n-InAs, $n = 2.5 \times 10^{16}$ cm^{-3} (Ref. [45]). (– – –, ——): calculated data. ●: data from mobility measurements, ○: data from SdH measurements

pendence of the mobility on the electron temperature. Therefore also no remarkable dependence of μ on E may be expected as long as $\eta \gg 20$ and this method will be unsuitable for determining electron temperatures. Figure 3 shows the field dependence of the mobility of n-InAs [45] at a lattice temperature of 4.2 K ($n = 2.5 \times 10^{16}$ cm^{-3}, $\eta = 37.5$). Changes in the mobility are apparent for fields in excess of 200 mV/cm where T_e is already at 8 K. Only the full dots in the $T_e(E)$ curve were determined from mobility measurements whereas the open dots are taken from Shubnikov de Haas measurements. The $T_e(E)$ curve also shows a steep increase between 0.2 and 0.4 V/cm. Hall measurements in magnetic fields of 500 G did not show any evidence for a two band conduction. Thus a change of n does not occur at 500 G in such high doped samples. The full and broken lines are based on a calculation using the electron temperature model (see Section 4.1).

2.2.2. Hot Electron Hall Effect and Magnetoresistance

The application of the magnetic field (B) superimposed to a strong electric field will cause a reduction of the rate of gain of energy due to the deflection of the carriers caused by the Lorentz force. Thus a magnetic field will give rise to a "cooling effect" if the mean carrier energy at a certain electric field is compared with and without application of B.

In the calculations, the magnetic field has to be included in the field term of the BE

$$(\partial f/\partial t)_{\text{field}} = -(e/\hbar)(E + c^{-1} v \times B) \, V_k f \, .$$

Approximate analytical solutions of the hot carrier problem are again restricted to acoustic phonon and ionized impurity scattering [10, 46–48].

In order to illustrate the influence of a magnetic field on high field transport in the electron temperature model we use the following definitions: in an isotropic material the Hall constant R_B is given by [44]

$$R_B = B^{-1} \, \sigma_{xy}(B) \, [\sigma_{xx}^2(B) + \sigma_{xy}^2(B)]^{-1}$$

for $B \| z$ direction. σ_{ij} are the components of the conductivity tensor which relate j to E and depend on B

$$j_i = \sum_j \sigma_{ij} E_j \, .$$

The transverse magnetoresistance $\varrho_{xx}(B)$ is given by

$$\varrho_{xx} = \sigma_{xx}(\sigma_{xx}^2 + \sigma_{xy}^2)^{-1} \, .$$

σ_{xx} and σ_{xy} are given by

$$\sigma_{xx} = (ne^2/m) \langle \tau(\varepsilon) (1 + \omega_c^2 \tau^2(\varepsilon))^{-1} \rangle \tag{30a}$$

$$\sigma_{xy} = -(ne^2 \omega_c/m) \langle \tau^2(\varepsilon) (1 + \omega_c^2 \tau^2(\varepsilon))^{-1} \rangle \, . \tag{30b}$$

In the limit of vanishing magnetic fields ($\omega_c \tau \ll 1$) the Hall constant R_B is thus given by

$$R_B = r/nec \qquad r = \langle \tau^2 \rangle / \langle \tau \rangle^2 \, . \tag{30c}$$

r is the Hall coefficient factor.

For the averaging

$$\langle h(\tau(\varepsilon/k_B T_e)) \rangle = \frac{\int\limits_0^\infty h(\tau(\varepsilon/k_B T_e)) (\varepsilon/k_B T_e)^{3/2} (\partial f_0/\partial(\varepsilon/k_B T_e)) \, d(\varepsilon/k_B T_e)}{\int\limits_0^\infty (\varepsilon/k_B T_e)^{3/2} (\partial f_0/\partial(\varepsilon/k_B T_e)) \, d(\varepsilon/k_B T_e)} \tag{30d}$$

has to be used.

In the energy balance equation σ_{xx} will enter and therefore cause a reduction of the power input from the electric field [47][7].

Equation (30c) demonstrates clearly that changes of the Hall constant will occur in an electric field even if the carrier concentration

---

[7] The expressions entering in the energy balance equations are different for a mobility and a resistivity measurement (see Ref. [47]).

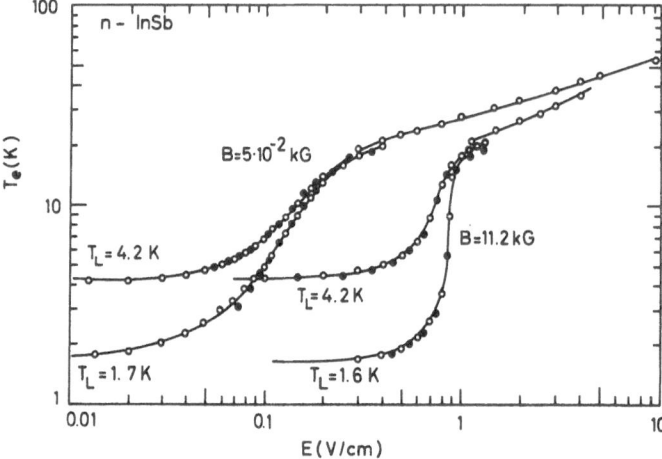

Fig. 4. Comparison of electron temperatures deduced from mobility (○) and Hall measurements (●). (After Miyazawa [25])

stays constant: r is an average of the relaxation time over the distribution function which change in high electric fields.

In high magnetic fields and at low temperatures ($\omega_c \tau_m \gg 1$, $\hbar\omega_c > k_B T$) the conduction or valence band will split into Landau levels whose spacing is $\hbar\omega_c$ for parabolic bands [49]. This quantization will also influence the scattering rates of the electrons with impurities and phonons [50].

All this complications make Hall measurements not very suitable for the determination of electron temperatures. Nevertheless it has been tried several times using n-InSb with carrier concentrations of about 10^{14} cm^{-3} at Helium temperatures [25, 32, 51, 52]. Miyazawa's results [25] for a sample with $n = 1 \times 10^{14}$ cm^{-3} are shown in Fig. 4. Electron temperatures are deduced either from Hall measurements and from transverse magnetoresistance measurements and yield up to $T_e = 20$ K the same temperature values.

The effect of the transverse magnetic field to "cool" the carriers is clearly demonstrated by comparing the curves for $B = 5 \times 10^{-2}$ kG and $B = 11.2$ kG. However at 11.2 kG, already $\omega_c \tau \gg 1$, and in addition appreciable magnetic freeze-out occurs. The observed $\sigma(E)$ characteristics on which the $T_e(E)$ curves are based, are therefore influenced by a change of the carrier concentration with increasing electric field.

All experimental investigations [25, 32, 51, 52] on n-InSb ($n \approx 10^{14}$ cm^{-3}) at liquid He temperatures show an extremum of the Hall constant R_B for fields about 0.15 V/cm. Whereas in [25, 51, 52] this extremum is interpreted as an indication for two band conduction, in [32] and [41] another explanation is given. Sandercock [32] has made $R_B(E)$ measurements in n-InSb with carrier concentrations of 0.47×10^{14}, 2.2×10^{14}, and 4.4×10^{14} cm^{-3} for $T = 2\text{--}20$ K. At a magnetic field of 160 G electric field dependent measurements up to 0.3 V/cm were made. The Fermi temperature ε_{F_0}/k_B of the samples is 4, 10.5, and 17 K. Since the Hall coefficient factor $r(\eta)$ for ionized impurity scattering is given by [44]

$$r_{0,I}(\eta) = \frac{3}{4} \frac{F_{7/2}(\eta) F_{1/2}(\eta) g^2(\bar{x}_1, \eta)}{F_2^2(\eta) g^2(\bar{x}_2, \eta)}. \tag{31}$$

r will tend to 1 for strong degeneracy, whereas for non-degenerate statistics $r_{0,I} = 1.93$. Thus with increasing electron temperature r increases. The maximum of R_B found in n-InSb is attributed to the onset of second momentum scattering process dominating ionized impurity scattering at high energies. Crandall [41] attributes the decrease of R_B at higher fields to acoustic phonon scattering.

Sandercock [32] has also performed a determination of $T_e(E)$ from his $R_B(E)$ measurements together with $\mu(E)$ measurements. He has pointed out that a comparison of $T_e(E)$ from these two methods should yield some information on the distribution function. Since μ involves only an averaging of τ whereas r involves an averaging of τ^2 over the distribution function, the electron temperature model is checked to an extent that r/μ is sensitive to changes in the distribution function. For electron temperatures up to 20 K, agreement between both measurements was obtained. The same result was got by Miyazawa [25].

On the other hand it was shown by Crandall [53, 54], that $R_B(E)$ experiments up to 30 V/cm in n-GaAs ($n = 3.5 \times 10^{15}$ cm^{-3}) at low temperatures (1.22–27 K), could not be explained on the basis of the electron temperature model. The magnetic fields were too small for Landau quantization to occur. A solution of the BE including impurity scattering and acoustic phonon scattering yielded good agreement of the observed $r(E)$ values with calculated ones [53, 54]. The electron temperature model could fairly well expalin the dependence of the mobility μ on E whereas $R_B(E)$ could not be explained in the experiments on n-GaAs.

Calculations for high magnetic fields ($\hbar\omega_c > k_B T_e$) for deformation potential scattering were carried out by Kazarinov and Skobov [55] who got for the $B \perp j$ configuration a MB distribution with $T_e = T \cdot (1 + \frac{1}{2}(cE/uB)^2)$, c being the velocity of light. Experiments by Fujisada et al. [56] on n-InSb at 77 K found qualitative agreement with the predictions of Kazarinov and Skobov [55] [8].

---

[8] Under the experimental conditions however polar optical phonon scattering will be important.

Experiments on the influence of high transverse and longitudinal magnetic fields up to 20 kG on the $j - E$ characteristics of n-InSb ($n = 1.7 \times 10^{14}$ cm^{-3}) at 1.8 and 4.2 K have also been performed by Kotera et al. [57]. In explaining their results Kotera et al. take into account quantized subband structures and associated change in scattering rates and are able to describe at least qualitatively the observed $j - E$ characteristics. The calculations are based on a diffusion of particles in energy space and yield electron energy distribution functions which are severely distorted from a MB distribution at the Landau level energies and above the optical phonon energy [58]. Unfortunately Landau level broadening was not taken into account, this may be the reason for the lack of quantitative agreement between theory and experiment.

It may be concluded that $r(E)$ measurements have only proved twice in n-InSb ([25, 32]) that they are capable of yielding electron temperatures in agreement with data from $\mu(E)$ measurements and that this method is restricted to very small magnetic fields.

2.2.3. Variation of the g-Factor of Conduction Electrons in a Non-Parabolic Band

Until now we have only considered parabolic bands. The nonparabolicity of the conduction band of n-InSb e.g. can be exploited for the determination of electron temperatures by measuring the field dependence of the gyromagnetic ratio or g-factor of the conduction electrons. It was already mentioned that the conduction band is splitted into Landau levels in high magnetic fields. Each Landau level will be again splitted in two levels, corresponding to the two possible spin orientations (Fig. 5). The splitting of these levels depends on the g-factor. In a nonparabolic band the effective g-factor is a function of the energy, it decreases with increasing energy. Thus the energy difference between spin-split levels decreases with increasing quantum number N.

For the Kane-model, applicable to InSb, the dependence of g on the energy will be given by [59]

$$g(\varepsilon) = g_0 \Delta^{-1} [\Delta \varepsilon_g (\Delta \varepsilon_g + \Delta)] \cdot [(\varepsilon + \Delta \varepsilon_g)^{-1} - (\varepsilon + \Delta \varepsilon_g + \Delta)^{-1}] \tag{32}$$

where g_0 is the g-factor of the bottom of the conduction band and Δ is the spin-orbit splitting energy and $\Delta \varepsilon_g$ is the gap energy.

The spin splitting measured in an electron spin resonance (ESR) experiment yields a g-factor averaged over the distribution function [60] and will thus depend on the electron temperature. Isaacson and Feher [61] have first reported, that in high electric fields a shift of the

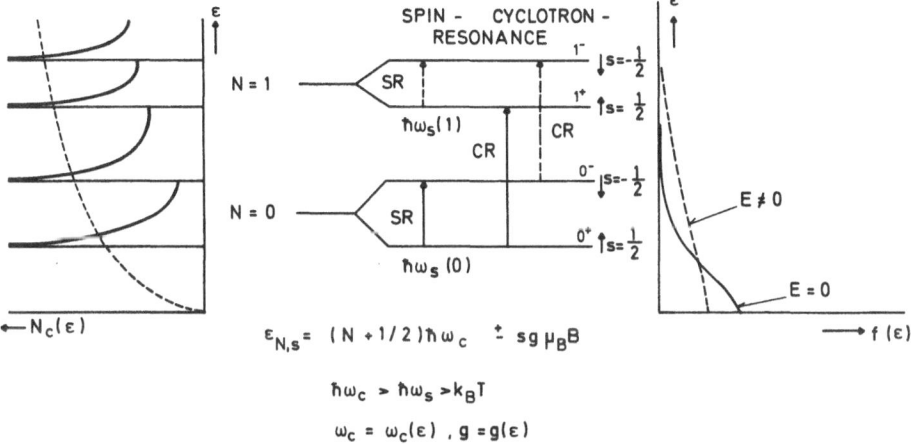

Fig. 5. Landau level and spin splitting. LHS: density of states for $B \neq 0$ (———) and for $B = 0$ (–––). RHS: change of $f(\varepsilon)$ with E indicating occupation of $N = 0$, $s = -1/2$ level (0^-) and $N = 1$, $s = 1/2$ (1^+) level with increasing E (–––). In spin resonance thus 1^+–1^- transitions and in cyclotron resonance 0^-–1^- transitions are possible for $E \neq 0$

ESR line will occur which can be interpreted in terms of an increasing electron temperature [9].

Guéron [62] has presented experimental details on the variation of the g-factor with electric field in n-InSb at a lattice temperature of 1.7 K. The sample was placed in a microwave cavity for 36 GHz and the electron spin resonance signal was detected by magnetic field modulation and lock-in detection (essentially the derivative of the resonance signal was detected). For a carrier concentration of $n = 1.4 \times 10^{14}$ cm^{-3} the Fermi level was already within the conduction band at 1.7 K. First the dependence of the resonance position on lattice temperature was measured which showed a linear decrease of the $|g|$-factor for temperatures approximately higher than 10 K. Later experiments by Kaplan and Konopka [60] showed that for InSb with a carrier concentration of 1×10^{14} cm^{-3} the decrease of the $|g|$-factor with temperature below 10 K ($T \ll T_F$) is also linear, in contrast to expectations based on the unperturbed band model and Fermi statistics. These authors suggested an impurity band and an increasing population of the conduction band

[9] Recently far infrared cyclotron resonance transitions 0^+–1^+, 0^-–1^-, 1^+–2^+ and 1^-–2^- in n-InSb occuring in high electric fields have been exploited for a determination of T_e assuming that the absorption in the cyclotron resonance line is proportional to the number of electrons in the initial Landau level subband. The electrons were assumed to be distributed among the subbands according to a MB distribution. See: Kobayashi, K. L. I., Otsuka, E.: Proc. Int. Conf. Physics of Semiconductors, Miasek, M. (Ed.). Warsaw: Polish Scientific Publishers 1972, p. 903.

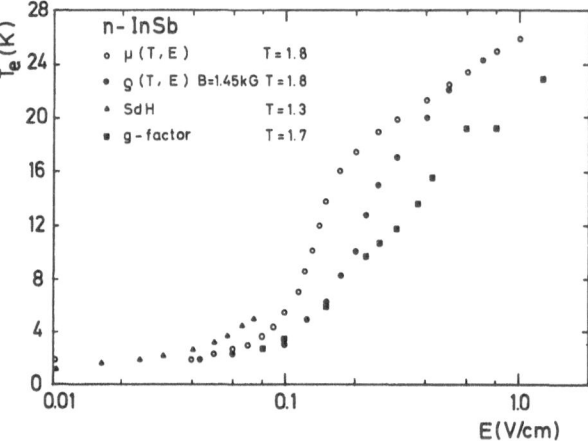

Fig. 6. $T_e(E)$ for n-InSb deduced from mobility ($\bigcirc$, $n = 1 \times 10^{14}$ cm^{-3}, Ref. [25]), magneto-resistance ($\bullet$, $n = 1 \times 10^{14}$ cm^{-3}, Ref. [25]), SdH effect ($\triangle$, $n = 1.7 \times 10^{15}$ cm^{-3}, Ref. [68]) and g-factor ($\blacksquare$ $n = 1.4 \times 10^{14}$ cm^{-3}, Ref. [62])

with lattice temperature to explain the linear behaviour of the g-factor at low temperatures.

Guéron deduced the increase of the electron temperature with electric field by observing the shift of the ESR line with electric field up to 2 V/cm at a lattice temperature of 1.7 K and comparing these data with the shift under ohmic conditions with lattice temperature. The heating electric field was a d.c. field. The orientation of the magnetic field (ca. 500 G) was perpendicular to the applied electric field. No dependence of the ohmic mobility on the magnetic field was observed, which might be due to the fact that the samples used were polycrystalline. Results of Guéron's experiments are presented in Fig. 6 and show that the $T_e(E)$ curve deduced from these measurements is shifted to higher values of E compared with data from mobility measurements. The reason for this shift may be the poor quality (polycristalline) of the samples in the g-factor experiment.

Later Konopka [63] has repeated ESR measurements in high quality single crystal n-InSb with the main objective of detecting beside the $0^+–0^-$ transition also $1^+–1^-$ transitions at higher T_e where the $N = 1^+$ Landau level becomes populated. The measurements were performed at 105 GHz and a lattice temperature of 1.6 K. The electrons were heated in the microwave field and the resonance experiment was performed with the microwave power level as a parameter (Fig. 7). The determination of the electron temperature assigned to the different curves was not explicitly described. It could be either calculated or

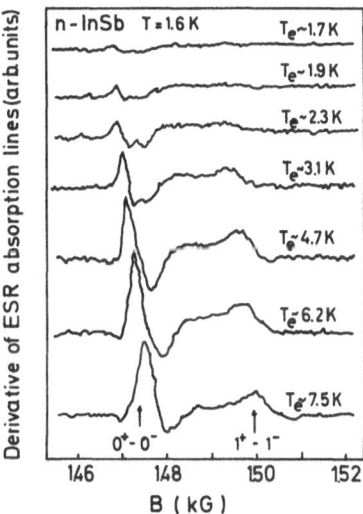

Fig. 7. Electron spin resonance of hot electrons in InSb. For $T_e > 3.1$ K, the $N = 1$ Landau level is populated. (After Ref. [63])

deduced by comparison with experiments at sufficiently low microwave power level. A formula equivalent to Eq. (32) [64] was used for a calculation of the g-factor, and again a difference between the experimental slope of $g(T_e)$ and that of the calculated $g(T_e)$ was found. The spin population difference for the $N = 0, 1$ levels as a function of the temperature was calculated and compared with the integrated amplitudes. By assuming the same transition probabilities for the $N = 0, 1$ levels, the same spin-lattice relaxation time, the comparison of the integrated amplitudes with the calculated spin-population at $T = 4.7$ K is in satisfactory agreement. No further analysis was made for other temperatures and a $T_e(E)$ or $T_e(P)$ curve was not given so that a comparison of this method with others is unfortunately not possible. In as far an impurity band or band tailing may have influenced the results is not discussed.

2.2.4. Hot-Electron Shubnikov de Haas Effect

The Shubnikov de Haas (SdH) effect is an oscillatory variation of the magnetoresistance in high magnetic fields, which is found in degenerate materials at low temperatures. Changes in the magnetic field result in a passage of the Landau levels through the Fermi energy. As a result oscillations in the magnetic susceptibility and in transport properties occur, which are periodic as a function of the reciprocal magnetic field.

The conditions necessary for the oscillatory effects to be observed are the following

$$\omega_c \tau \gg 1 \qquad \hbar\omega_c > k_B T_e \qquad \varepsilon_F > \hbar\omega_c \tag{33}$$

where $\omega_c = eB/mc$, for parabolic bands.

The theory of the transverse and the longitudinal magneto-resistance yields for the change of resistivity [65]

$$\Delta\varrho/\varrho_0 = a \sum_{s=1}^{\infty} [b_s \cos(2\pi s\varepsilon_F/\hbar\omega_c - \tfrac{1}{4}\pi) + R] \tag{34}$$

with

$$b_s = (-1)^s s^{-1/2} (\hbar\omega_c/2\varepsilon_F)^{1/2} \frac{2\pi^2 s k_B T/\hbar\omega_c}{\sinh(2\pi^2 s k_B T/\hbar\omega_c)} \cos(\pi\nu s) e^{-2\pi^2 k_B T_D/\hbar\omega_c}$$

where ϱ_0 is the zero field resistivity, ν the ratio of spin-splitting to Landau level spacing, and T_D the non-thermal broadening temperature (Dingle-temperature [66]). The term R represents an additional series which is generally much smaller than b_s and vanishes in the longitudinal case. a is equal to 2.5 or 1 for the transverse or longitudinal configuration respectively. Retaining only the first term in the series, the ratio of the amplitudes A at two different electron temperatures is given by

$$A(T_{e,1})/A(T_{e,2}) = \chi_1 \sinh\chi_2/\chi_2 \sinh\chi_1 \tag{35}$$

where

$$\chi_i = 2\pi^2 k_B T_{e,i}/\hbar\omega_c .$$

Thus the amplitudes are damped with decreasing magnetic field and strongly depend on the temperature which enters in the carrier distribution function.

Whereas in non-degenerate material the variation of the mobility with electric field yields information on the electron temperature, for highly doped materials ($\eta > 10$), this method cannot be applied. In the limit of strong degeneracy the mobility due to ionized impurity scattering in a semiconductor does not depend on the lattice temperature and therefore also not on the electron temperature (see p. 18). For this reason a method which determines directly the electron temperature is more suitable to study hot electron effects in degenerate semiconductors than the observation of the variation of the conductivity with field. The above outlined sensitivity of quantum oscillations like the SdH effect on the electron temperature makes this method useful for a study of hot electron effects.

The first experiment concerning the influences of electric fields of the order of 100 mV/cm on the SdH oscillations was performed by Komatsubara [67] in n-InSb. He observed in addition to the shift of the extrema of the resistance oscillations to higher magnetic fields, a decrease of the oscillatory amplitudes with increasing electric field. The shift was explained by a Stark effect, apparently assuming that the Fermi energy was constant as the electric field changed. Later Isaacson and Bridges [68] repeated these experiments in n-InSb and explained the shift as caused by a temperature and magnetic field dependent non-oscillatory magnetoresistance superimposed to the SdH oscillations. The damping of the oscillations was attributed to a change of the mean carrier energy or the electron temperature with applied electric field. The electron temperature was deduced from a comparison of the dependence of the amplitudes on the lattice temperature at a low value of the electric field strength with the dependence of the amplitudes on the electric field strength at a fixed low value of the lattice temperature.

Despite the high carrier concentration and the high mobility of the samples under investigation several authors [67–70] applied electric fields permanently so that even at low fields a rise of lattice temperature could not be excluded because of the thermal properties at low temperatures (see Ref. [71]). Bauer [72] has used a pulsed d.c. technique in measuring SdH oscillations in degenerate InAs thus avoiding a change of lattice temperature. Detailed measurements of the hot electron SdH effect in n-InAs, n-InSb, n-GaSb have later been made by Bauer and Kahlert [73–77].

The experimental arrangement is shown in Fig. 8. Voltage pulses of a duration between 0.5 and 2 µs were produced by a delay line pulse generator. The sample was placed in a temperature variable insert or directly immersed in liquid Helium, within a superconducting coil. In order to observe any variation of the conductivity with electric field, with magnetic field and with time after the application of the pulse two sampling oscilloscopes were used. For small signals additional amplification with pulse amplifiers was necessary prior to feeding the signals to the oscilloscopes. Using an $x - y$ recorder and the field sweep of the superconducting coil direct plots of the magnetoresistance vs. B were obtained.

Figure 9 shows a comparison of SdH oscillations in n-InAs having a carrier concentration of 2.5×10^{16} cm^{-3} and a mobility $\mu_0 = 25000$ cm^2/ Vs at 4.2 K. In order to evaluate electron temperatures from this comparison, the Dingle temperature has to be constant. In the electric field range up to 300 mV/cm the mobility was constant within 2% so that it was reasonable to assume that also the Dingle temperature did not vary. Furthermore, after having assigned an electron temperature to a

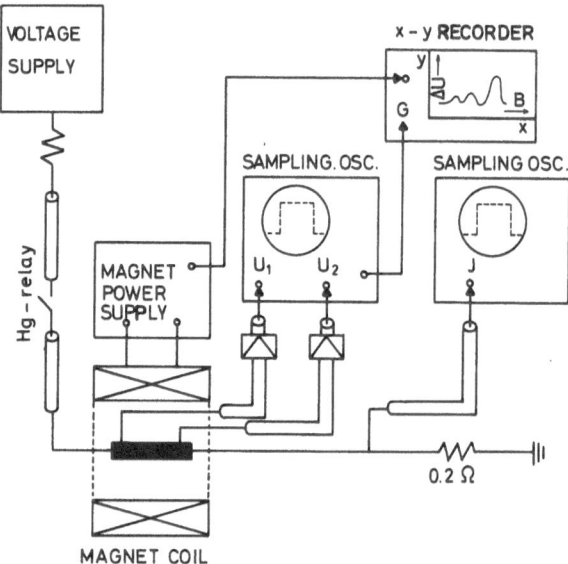

Fig. 8. Experimental arrangement for pulsed d.c. SdH measurements

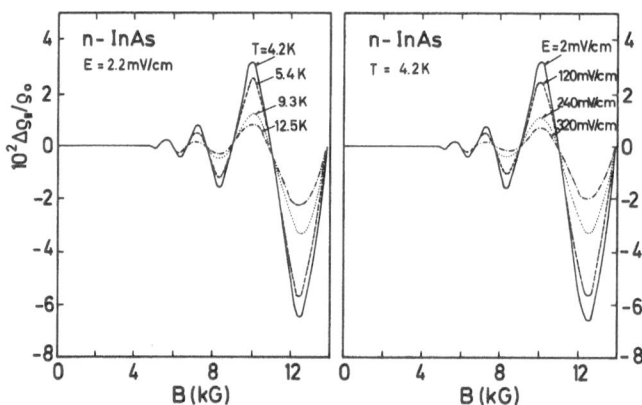

Fig. 9. Comparison between the influence of increasing lattice temperature on the amplitudes of the SdH oscillations with the influence of electric fields (2–320 mV/cm) for n-InAs. $B \parallel j$. (After Ref. [74])

particular curve, it was possible to deduce the Dingle temperature from the oscillations and a consistent result was obtained: the electron temperatures were determined correctly and did not depend on the magnetic field and the Dingle temperature did also not vary with applied electric field, both within the experimental error of a few percent. Calculations based on the Eqs. (34), (35) for different temperatures were also

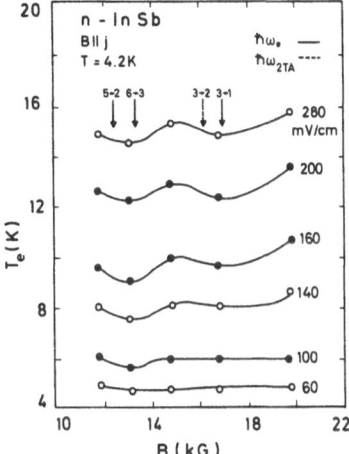

Fig. 10. Variation of T_e with magnetic field for n-InSb ($n = 6.9 \times 10^{16}$ cm^{-3}) deduced from SdH measurements: $E = 60$–280 mV/cm. Arrows indicate magnetic field positions for resonant emission of $2TA$ phonons (10.3 meV) and optical phonons (23,9 meV)

compared with the experimental oscillations and a good agreement was obtained.

As a further check the electric field dependence of the mobility was measured at higher electric fields up to 10 V/cm. The strong degeneracy ($\eta = 37.5$ at 4.2 K) was gradually removed by the field so that mobility changes were observable for $E > 200$ mV/cm. Electron temperatures were deduced from a comparison with ohmic data of the temperature dependence of the mobility. In a region of electric fields where data obtained, both from the SdH oscillations and from the mobility measurements, electron temperatures deduced from both methods are in satisfactory agreement (see Fig. 3). The sample concentration was chosen in such a way to allow this overlap experiment. Whereas at low T_e it was sufficiently degenerate for SdH measurements, with increasing T_e the degeneracy is gradually removed and field and temperature dependent changes of the mobility are detectable.

For the evaluation of the electron temperature only oscillatory amplitudes were used, which did not change the resistivity by more than about 6% to ensure approximately constant power input from the electric field at varying B. Thus we have restricted the analysis to relatively small magnetic fields and have not involved $N = 0$ and $N = 1$ Landau levels. Nevertheless it is astonishing that no magnetic field dependence of T_e was found. In a subsequent experiment on n-InSb we have made a very extensive study on this dependence of the electron temperature on the magnetic field and the results are shown in Fig. 10.

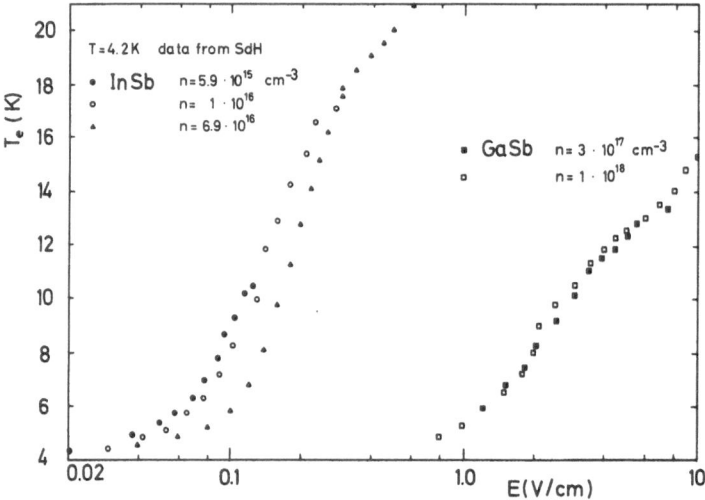

Fig. 11. $T_e(E)$ for n-InSb and n-GaSb deduced from SdH measurements (Ref. [75, 76])

We have used higher doped material in order to investigate more Landau levels. The electron temperatures determined at extremal positions change somewhat. These changes are not only a manifestation of the magnetic field dependent oscillatory power input from the electric field resulting from the experimental constant current conditions. More precise calculations of the energy loss rate [78] indicate that a kind of "resonance cooling" may occur if at a certain magnetic field the separation of Landau levels equals the polar optical phonon energy or two times the transverse acoustic phonon energy of the X point of the Brillouin-zone [79]. The arrows in Fig. 10 indicate the magnetic field positions for different transitions calculated by taking into account the non-parabolicity of the conduction band.

Since the oscillations of T_e with magnetic field are small, an averaged value of T_e can be related to each electric field E, regardless of the magnetic field. Calculations based on the electron temperature concept using a Fermi distribution function are in good agreement with the experiment. The reasons for this behaviour are the following: (i) in all experiments the Fermi velocity $v_F > 100\,v_d$, due to the restriction to small electric fields; (ii) the magnetic field is restricted such, that the non-oscillatory and oscillatory magnetoresistance do only change the zero field resistance by a few percent.

The increase of T_e with E deduced from SdH measurements is shown in Fig. 11 for n-InSb and n-GaSb for different carrier concentrations. The difference in the range of fields which cause an appreciable increase of T_e with E, is caused by the different carrier concentrations and material parameters like mobility, effective mass, and coupling constants. The curves can theoretically be described by taking into account ionized impurity scattering for the momentum exchange with the electrons and acoustic phonon and optical phonon scattering for the energy loss processes [75, 76].

Although there is a rapid increase of T_e with E observed in Fig. 11, one should bear in mind that the mean energy defined according to Eq. (27) will for a strongly degenerate system ($\eta \gtrsim 20$) only be a weak function of T_e and can be expanded according to Wilson [80]

$$\langle \varepsilon \rangle = \tfrac{3}{5} \varepsilon_{F_0} \{1 + \tfrac{5}{12} \pi^2 (k_B T_e / \varepsilon_{F_0})^2 - \tfrac{1}{16} \pi^4 (k_B T_e / \varepsilon_{F_0})^4 - \cdots \} \tag{36}$$

where ε_{F_0} is the Fermi energy of $T = 0$.

Hence the mean energy does not increase strongly with T_e. The quantity which demonstrates this most clearly is the specific heat of the electrons $d\varepsilon/dT = c_h$, which shows at low temperatures for $\eta \gg 20$ approximately a linear increase with T and finally tends at high temperatures, when the degeneracy is removed, to its temperature independent value of $\tfrac{3}{2} k_B$ for Maxwell-Boltzmann statistics.

2.2.5. Time Resolved Observation of Increasing Electron Temperatures

Until 1962, when Schmidt-Tiedemann published his review on hot electrons [6] mainly Ge and Si had been investigated. These materials exhibit momentum and energy relaxation times of the order of 10^{-13}–10^{-11} s for temperatures of 77 K and higher. Therefore only indirect methods like the field dependence of the microwave conductivity yielded information on these times if the frequency of the microwaves is of the order of the reciprocal of the relevant relaxation time [6].

At lower temperatures it is expected that the energy relaxation time will increase significantly due to the fact that only few electrons in the high energy tail of the distribution function will be able to emit optical or high energy acoustic phonons. Also the energy dissipation by long wave length acoustic phonon scattering will be small. Early experiments on n-InSb at helium temperatures by Sladek [35] confirmed these expectations and showed that fields of the order of several 100 mV/cm produce appreciable carrier heating, indicating a large τ_ε.

For a discussion of the time dependence, we start with the time dependent energy balance equation [Eq. (25)]. Generally no analytical solution of this equation can be given and only numerical solutions can

be obtained[10]. However in order to demonstrate the basic principle
we assume that μ and τ_ε are independent of ε. Then

$$\langle \varepsilon \rangle = \varepsilon_L + \tau_\varepsilon e \mu E^2 [1 - \exp(-t/\tau_\varepsilon)]. \tag{37}$$

Thus the mean energy is an increasing function of time, its final value
determined by τ_ε and $e\mu E^2$. For $t > 5\tau_\varepsilon$ $\langle \varepsilon \rangle$ has reached approximately
a saturation value within 1 %.

Due to the fact that the mobility for almost all scattering mechanisms
is a function of the electron energy, a time dependent change of the mean
energy will result in a time dependent mobility variation. A stationary
state will be achieved for times long compared to the energy relaxation
time, being usually longer than the momentum relaxation time.

Peskett and Rollin [34] have first carried out a direct observation
of the influence of τ_ε on the voltage pulse shape in n-InSb at 4.2 K under
constant current conditions. For an electric field of the order of 100 mV/
cm they observed a time dependent voltage drop in a μs time scale
corresponding to a time dependent decrease of the resistivity. If a second
pulse was applied at 0.4 μs after the end of the first one, the initial peak
of this second pulse was not as high as that of the first pulse. In a time
of 0.4 μs the electron system had not been able to return to its equilibrium
with the lattice and thus the initial mobility was already higher than in
equilibrium with the lattice at 4.2 K. From this mobility change an energy
relaxation time of 3.3×10^{-7} s was deduced, by the authors, using a
simple $\mu \sim T^{3/2}$ dependence.

Using a sampling oscilloscope having a rise time below 1 ns, Sandercock [31] was able to deduce directly from the voltage waveforms
energy relaxation times between 11.6 ns and 3.5 ns for temperatures
between 14.6 and 20.4 K in n-InSb with a carrier concentration of
6.9×10^{13} cm^{-3}. Due to the energy dependence of τ_ε and τ_m in n-InSb
Sandercock's method has some disadvantages since no clear exponential
behaviour according to Eq. (37) can be expected. To overcome this
difficulty Maneval et al. [26] have used a very elegant technique which is
based essentially on applying a double step voltage pulse with $\Delta E \ll E$
and observing the current response (insert of Fig. 12).

The variation of the current, following the voltage steps, consists
of an initial jump determined by the momentum relaxation time (of the
order of 10^{-12} s) and followed by a slow further variation of the current,
since the mean energy increases as determined by the energy relaxation
time. The mobility increases due to the dominating ionized impurity
for τ_m which causes approximately a $\tau_m \sim \varepsilon^{3/2}$ dependence.

---

[10] For the "warm carrier" region an iterative solution of the coupled momentum and
energy balance equation can be given. See Seeger, K., Hess, K. F.: Z. Phys. **237**, 252 (1970).

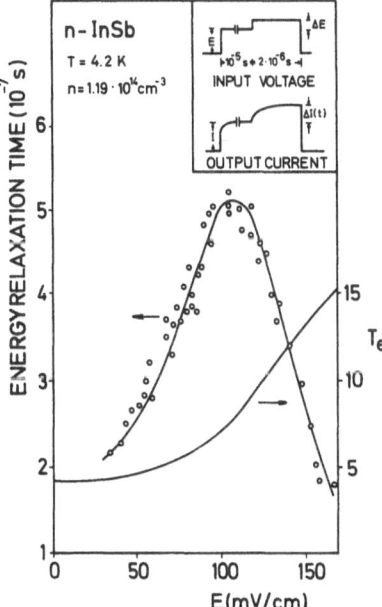

Fig. 12. τ_ε and T_e vs electric field in n-InSb. Insert: double step voltage pulse applied to the sample and output current. (After Ref. [26])

For a small step variation ΔE the energy variation $\Delta\varepsilon(t)$ is given in a linear approximation by

$$\Delta\varepsilon(t) = 2e\mu E \Delta E \tau_\varepsilon (1 - \exp(-t/\tau_\varepsilon)).$$

The step variation ΔE on the top of a voltage pulse of height E followed after a time long enough that already stationary conditions prevailed $(10^{-5}\,\text{s} > 5\tau_\varepsilon)$. From the time dependence of the current rise τ_ε can be measured and can be determined as a function of the applied field E. Then the average energy can be calculated from

$$\langle\varepsilon(E)\rangle - \varepsilon(0) = (2/n) \int_0^E j(E)\,\tau_\varepsilon(E)\,\mathrm{d}E$$

and thus the energy dependence of μ is also determined. The average energy can be related to the electron temperature by using Eq. (27).

In the experimental arrangement a well defined constant voltage source with a short risetime was built in a $50\,\Omega$ coaxial system. The current, measured as a voltage drop on a disc resistor, was amplified with a wideband amplifier and fed into a sampling oscilloscope. The output of the oscilloscope was connected to a multichanal analyser which essentially extracted the small signal (due to the small ΔE) from the

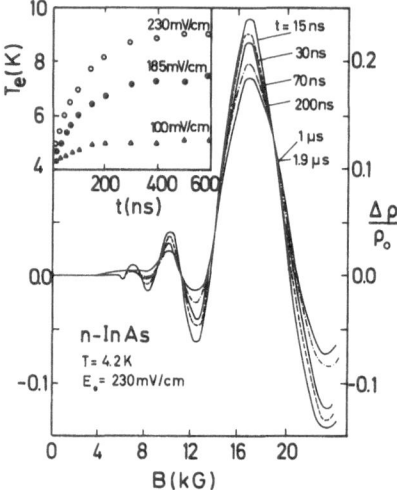

Fig. 13. Time dependence of the SdH oscillations in *n*-InAs for $E = 230$ mV/cm. Insert: increase of T_e with time for three electric fields ($T = 4.2$ K) [74]

noise. The bias field E was of the order of 40–400 mV/cm while ΔE was 5% of E. This small step was necessary in order to use the linear approximation in the solution of the time dependent energy balance equation [Eq. (38)].

Results for a sample with $n = 1.19 \times 10^{14}$ cm^{-3} and a lattice temperature $T = 4.2$ K are presented in Fig. 12 for electric fields up to 170 mV/cm. Both the increase of T_e with field and the dependence of τ_ε on E is demonstrated. Maneval et al. have also determined in the same sample the change of T_e with E by comparing field dependent mobility data with the increase of the ohmic mobility with lattice temperature ("static measurements"). This independent method yielded excellent agreement with the results of the "dynamic measurement".

The methods just described are restricted to weakly degenerate semiconductors. In a strong degenerate semiconductor ($\eta > 10$) the mobility changes with increasing electron temperature will be too small for direct time dependent observations but the SdH effect can be used. The time dependent increase of T_e after applying the electric field will cause a time dependent decrease of the SdH amplitudes. This method was first used by Bauer in *n*-InAs [72] and yielded energy relaxation times of the order of 10^{-7} s at $T = 4.2$ K and 200 mV/cm. The experimental arrangement is the same as already described in Section 2.2.4. A resolution of a few ns could be achieved. Figure 13 shows data on the damping of SdH oscillations for a time between 15 ns and 1.9 μs after application of a pulse corresponding to 230 mV/cm. This time dependent decrease of the oscillation amplitudes saturates after a few hundred ns. From the damping of the oscillations the increase of T_e with t as shown in the insert was deduced. Also shown is the $T_e(t)$ dependence for 185 and 100 mV/cm. By a numerical solution of the time dependent energy balance equation taking into account the energy dependence of τ_ε this time dependent increase of $T_e(t)$ can be explained quantitatively [74]. Similar experiments were also performed with degenerate n-InSb at 4.2 K with doping levels from $5.9 \times 10^{15} - 1 \times 10^{16}$ cm^{-3} [81].

The time dependent measurements in non degenerate or weakly degenerate materials have shown very nicely both relaxation times involved in a hot electron experiment. The SdH measurements have in addition demonstrated that in a degenerate semiconductor a method, which directly depends on the electron temperature is capable of detecting quite small changes in the mean energy of the electrons.

2.2.6. Investigations of the Noise Temperature

Whereas until now we were mainly concerned with methods for the determination of T_e at helium temperatures and comparatively small electric fields, Sections 2.2.6–2.2.9 are devoted to experiments in fields of the order of 0.1 kV/cm–2 kV/cm and lattice temperatures of 77 K and higher.

A method which yields information not only on the mean carrier energy but also on deviations from a MB distribution was proposed by Erlbach and Gunn [82, 83]. The theory for this experiment was developed by Price [84, 85]. In these investigations the hot carrier induced "noise" was determined. As under ohmic conditions, random fluctuations of the thermal velocity of the carriers will give rise to electrical noise. In analogy to Nyquist's formula [84] this noise can be characterized by a noise temperature T_n, where $k_B T_n$ is the available noise power per bandwidth. It may be expected that with increasing electric field this noise will rise and will give information on the carrier distribution.

However, beside the thermal noise there will be a convective noise due to fluctuations of the drift of the carriers, intervalley noise due to intervalley scattering in many valley semiconductors, surface trapping and generation-recombination noise. The latter source of noise is eliminated by measuring in the 30–500 MHz frequency range. In early experiments with n-Ge, also the convective and intervalley noise were eliminated as the noise voltage was picked up transverse to the applied electric field and current direction ([110]) in a [1$\bar{1}$0] direction.

The noise temperature T_n measured transverse to the applied field depends on the following averages over the actual distribution function [82, 83]

$$T_n = (e/k_B) \langle v_x^2 \tau \rangle / \mu_x' = T_n' \bar{\mu}_z / \mu_x' \tag{38}$$

where

$$\mu_x' = \partial \langle v_x \rangle / \partial E_x |_{E = E_z}$$

and

$$\bar{\mu}_z = \langle v_z \rangle / E_z .$$

For a MB distribution and spherical bands $T_n = T_n' = T_e$. A simultaneous measurement of the drift velocity parallel to the applied field and of the noise temperature should yield an information on the electron distribution function at least in the field region where only acoustic phonon scattering is important [82, 83].

Hart [86], Bareikis et al. [87] and Nougier et al. [88] later made investigations of the noise temperature parallel as well as perpendicular to the applied field. In all experimental set-ups the noise generated by the sample is compared with some standard noise source as can be seen e.g. in Hart's apparatus (Fig. 14). Bareikis et al. [87] used a microwave technique to make noise measurements in n- and p-type Ge and Si (see also [9]).

Only for a MB distribution and spherical bands T_n will represent T_e. Otherwise the connection between these two quantities is by no means simple. A theoretical analysis can be found in [86, 88]. In addition to theoretical difficulties due to the various sources contributing to noise there are also experimental ones as evidenced by the results got by different authors. Nougier et al. [88] have pointed out the delicate experimental problems associated with contact noise and the temperature dependence of the noise temperature.

Recently Monte Carlo calculations of hot-electron noise applicable to n-InSb have been performed [88a] which show that convective noise may be comparable to thermal noise in magnitude.

Both recent theoretical and experimental investigations have shown the limits of the use of the electron temperature concept and the difficulties in interpretation of experimental noise temperature results.

2.2.7. Thermoelectric Power of Hot Electrons

When a temperature gradient exists along a semiconductor sample, an open circuit voltage is built up. This so called Seebeck effect [89] is caused by a spatial variation of the carrier distribution function.

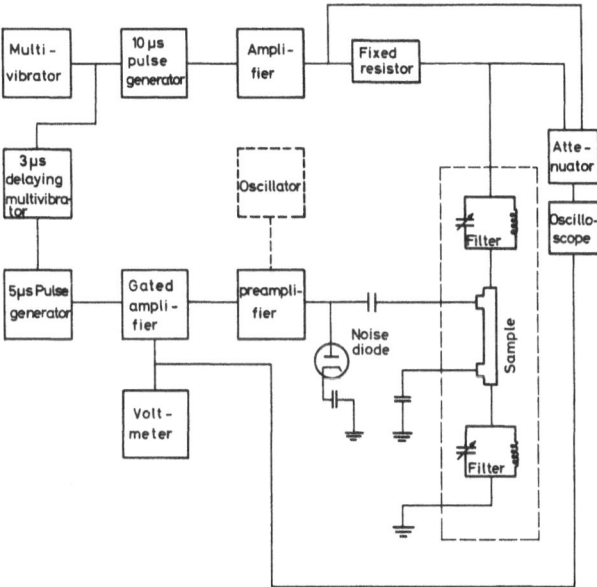

Fig. 14. Experimental arrangement for measurement of hot electron noise temperatures. (After Ref. [86])

A gradient in the carrier distribution function can also be caused by a gradient of the electric field. Then also a field gradient voltage or thermo-electric voltage will be established [90]. We only present a brief discussion since this effect has been reviewed in Ref. [9, 10][11].

In early d.c. experiments [91–93] it was hoped to deduce electron temperatures from the measurements of the field gradient voltage using the appropriate expression for the thermopower [90] and re-placing T by T_e. In order to establish a field gradient T-shaped samples have been used, so that in the middle of the sample the electric field exhibited a large gradient. However, due to the fact that the thermo-electric voltage is orders of magnitude smaller than the applied voltage, experimental difficulties in extracting the thermovoltage occured.

Later, by using high microwave fields[12], a gradient in carrier heating was obtained using special geometrical sample configurations [94–97] with contacts for picking up a d.c. thermoelectric voltage. However, just the contact regions which are necessary for the effect to build up, make it difficult to interpret quantitatively the observed d.c. voltage and its

[11] Another process based on the existence of a gradient in junctions, the hot carrier diffusion, has been exploited for the determination of electron temperatures in elemental semiconductors like Si (see e.g. Seeger, K., Ref. [22], p. 161).
[12] $\omega\tau_m < 1$, $\omega\tau_\varepsilon < 1$.

dependence on a.c. field strength. If both contacts were at the same T_e no effect would be measurable. If there is a gradient near a contact region however, the properties of the transition region will be of importance for the total thermoelectric d.c. voltage [95].

Experiments were performed with n- and p-Ge [94, 95] and with n-Si [96]. A summary of the work by Pozhela and coworkers on Ge and Si is found in Ref. [97] where also sources of experimental errors are discussed. n-CdTe was investigated by Alekseenko and Veinger [97 a]. The theoretical foundations for analyzing the microwave measurements were made by Conwell [10] and nonparabolicity effects were considered in Ref. [97 b].

Conwell [95] has pointed out that a comparison of measurements of the field gradient voltage with her theory is difficult due to the above mentioned contact effects and that therefore measurements of the field gradient voltage cannot quantitatively be interpreted.

2.2.8. Burstein-Shift of the Optical Absorption Edge

In this and in the following section we present optical methods for the experimental determination of T_e. First we discuss a method which is applicable to degenerate semiconductors.

Whereas in a non degenerate direct semiconductor the edge of the fundamental absorption is determined by the band gap energy $\Delta \varepsilon_g$, in a degenerate semiconductor this edge will be shifted to higher energies. If we consider a n-type semiconductor (Fig. 15) this shift is caused by the fact that in the vicinity of $k = 0$ no absorption is possible since all conduction band states are occupied. At an energy $4 k_B T_e$ below the Fermi-level about 98% of the states are filled and therefore the absorption will only take place for energies higher than this threshold. The shift ε_B [98]

$$\varepsilon_B = (\Delta \varepsilon_g)_{\text{opt}} - \Delta \varepsilon_g = (1 + m_c/m_v)(\varepsilon_F - 4 k_B T_e) \tag{39}$$

is called Burstein-shift. m_c and m_v are the effective masses in the conduction and valence band respectively, both assumed to be parabolic.

Shur [99] has pointed out that the Burstein-shift (BS) may be influenced by an external electric field. If the electrons are heated up, the distribution function becomes smoothed, the Fermi-energy will be reduced and therefore the absorption spectrum will alter. It will be possible to exploit this effect for a determination of the electron temperature as long as $\varepsilon_F - 4 k_B T_e > 0$. The temperature dependence of the Fermi-energy which has to be known can be determined from measurements of the Burstein-shift as a function of lattice temperature. This

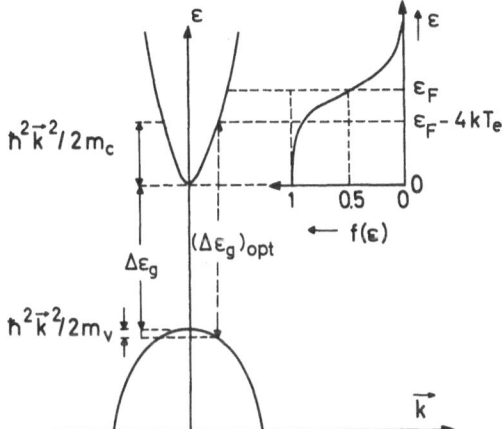

Fig. 15. Burstein-shift of the optical absorption edge in degenerate n-type semiconductors. (After Ref. [195])

can be done since the shift of Fermi-energy resulting from a variation either of the electron temperature or the lattice temperature is the same. The method is of course restricted to semiconductors where the interband transitions are dominated by direct ones and involve no emission or absorption of phonons.

Another influence of the electric field on the absorption edge, the Franz Keldysh effect [100, 101] will be several orders of magnitude smaller and hence will not interfere with the BS measurements if the range of electric fields is restricted to several 100 V/cm.

Heinrich and Jantsch [102] have made the first experimental investigation of this effect. They have used n-GaSb with a carrier concentration of 3.5×10^{17} cm^{-3} at a temperature of 77 K. The Fermi-energy was calculated by a two band model, taking into account besides the Γ-minimum also occupation of the L_1 minima. For this concentration $\varepsilon_F - 4 k_B T_e$ turns out to be 8.5 meV at 77 K. This value was experimentally checked by a comparison with the absorption spectrum of a sample having a concentration of 6.8×10^{16} cm^{-3} in which $\varepsilon_F - 4 k_B T_e < 0$ at 77 K and therefore no BS occurs. Experimentally a BS of 11 meV was deduced for the higher doped sample.

In the high field experiments polychromatic light has been used[13]. For a shift of the band edge by an amount of $\Delta \varepsilon$ the change in detector signal is given by

$$\Delta U = \int_0^\infty [T(\varepsilon + \Delta \varepsilon) - T(\varepsilon)]\, S(\varepsilon)\, d\varepsilon$$

[13] The range of photon energies was limited by the spectral sensitivity of the Ge-detector and by the properties of the samples to 0.65–0.85 eV.

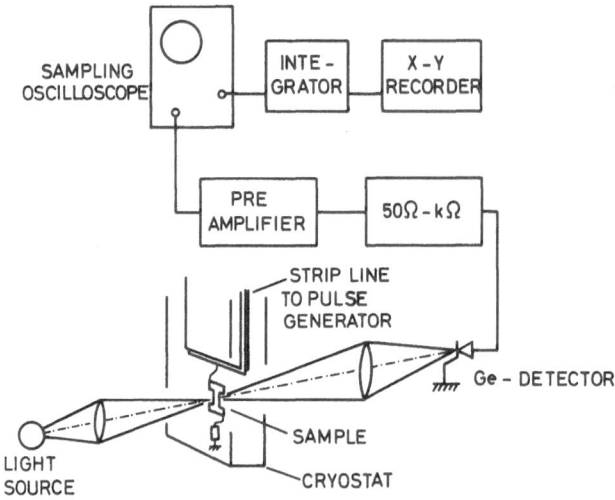

Fig. 16. Experimental setup for measuring the electric field dependence of Burstein-shift
(Ref. [103])

where $T(\varepsilon)$ being the transmittivity without shift and $S(\varepsilon)$ the spectral
sensitivity of the detector. The integrand may be expanded in a Taylor
series

$$\Delta U = \Delta\varepsilon \int\limits_0^\infty [\partial T/\partial\varepsilon + \tfrac{1}{2}(\partial^2 T/\partial\varepsilon^2)\,\Delta\varepsilon + \cdots]\,S(\varepsilon)\,\mathrm{d}\varepsilon.$$

Under the experimental conditions the second and higher order terms
could be neglected.

All data were related to the saturation value ΔU_S which is characterized by an electron temperature T_e such that

$$\varepsilon_F(T_e) - 4k_B T_e = 0.$$

The shift $\Delta\varepsilon$ at a certain temperature is given by

$$\Delta\varepsilon = (1 + m_c/m_v)\,[(\varepsilon_{F_0} - 4k_B T) - (\varepsilon_F - 4k_B T_e)]$$

so that

$$\frac{\varepsilon_F(T_e) - 4k_B T_e}{\varepsilon_F(T) - 4k_B T} = 1 - \Delta U/\Delta U_S. \tag{40}$$

The experimental arrangement is shown in Fig. 16 [103]. The change of
transmission due to an applied electric field was measured. Due to the
small sample resistance a pulse generator with an impedance of $1\,\Omega$
was used. The samples were bridge shaped to avoid contact effects.

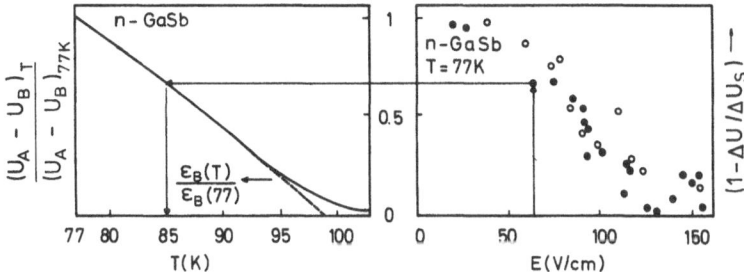

Fig. 17. Comparison of lattice temperature and electric field induced change of the optical absorption edge. (After Ref. [102])

Light from a tungsten filament was focused on the bar shaped part of the sample and the transmitted light focused on a Ge detector. An impedance transformer was used to match the detector output to the $50\,\Omega$ input of a preamplifier which was connected to a sampling oscilloscope. The rise time of the detector preamplifier system was about $0.4\,\mu s$.

In order to determine the temperature dependence of the Fermi-energy ε_F the temperature dependence of the transmission was measured between 83 and 105 K without applying an electric field. Due to the fact that GaSb shows a temperature dependence of the energy gap in this range which was not sufficiently exact measured previously, measurements on both the degenerate and the non degenerate material were performed. The changes in transmission on the non-degenerate (B) sample with temperature were attributed to the changing energy gap. Above 100 K when the BS of the degenerate (A) material vanishes both signals depended in the same way on temperature.

The BS as a function of lattice temperature was now obtained from the difference of the detector signals U_A-U_B of the samples A and B. In Fig. 17 these data normalized to U_A-U_B at 77 K are plotted vs lattice temperature. The right hand side of this figure shows $(1 - \Delta U/\Delta U_S)$, which is proportional to the BS, vs electric field. On the left hand side the dashed curve presents a calculation of the BS also normalized to 77 K.

The dependence of the electron temperature on the electric field can now be determined in a straight forward way as indicated in Fig. 17.

Figure 18 shows the results of this procedure together with a calculation based on an electron temperature model. For this calculation a two band model has been taken in account, assuming the same T_e in both the Γ and the L valleys and an energy relaxation time τ_ε of 1.6×10^{-11} sec [102]. The assumption of a constant τ_ε for fields up to 150 V/cm has been justified by analysing $j-E$ measurements and microwave measurements [104]. In this range of electric fields τ_ε changes less

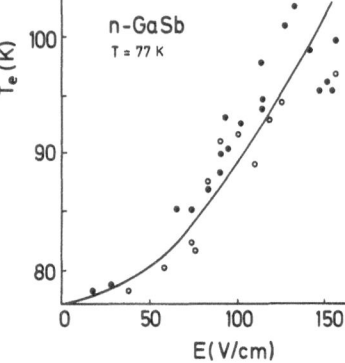

Fig. 18. $T_e(E)$ for n-GaSb ($n = 3.5 \times 10^{17}$ cm^{-3}) at $T = 77$ K, $\bigcirc$, $\bullet$: experimental data, (———): calculated data. (After Ref. [102])

than 10%. The reported results show that Burstein shift measurements are a valuable tool for the determination of electron temperatures in degenerate semiconductors. A refinement of this technique as discussed in Section 3.2.4 is even capable of determining energy distribution functions of hot carriers in degenerate semiconductors.

2.2.9. Hot Electron Faraday Effect and Birefrigence

In this section two methods are described which yield in principle information on $T_e(E)$ but have until now been mainly undertaken to get information on the redistribution of carriers among different conduction band valleys in high electric fields.

The Faraday effect is the rotation of the plane of polarization of linear polarized light for a propagation direction parallel to the applied magnetic field. Only the contribution from free electrons is of interest here and the Faraday rotation ϑ is given by the different velocities for left and right hand circularly polarized light [105].

Depending on the circular frequency of the electromagnetic radiation ω either a condition $\omega\tau \gg 1$ or $\omega\tau \ll 1$ can be established. In the high frequency limit and for weak magnetic fields ($\omega_c\tau \ll 1$) ϑ is approximated by

$$\vartheta = \frac{2\pi e^3 n B d'}{n_r c^2 m^2 \omega^2} \tag{41}$$

where n_r is the refractive index and d' the thickness of the sample in propagation direction. In this limit the Faraday angle will only depend on the effective mass and is independent of the relaxation time. In the limit $\omega\tau \ll 1$ and $\omega_c\tau \ll 1$ the rotation ϑ will depend on τ^2 [106].

$$\vartheta = \frac{2\pi e^3 n B d' \tau^2}{n_r c^2 m^2}. \qquad (42)$$

Thus hot electron effects will show up in the Faraday rotation either:

a) due to a change of the effective mass with E (nonparabolicity or intervalley transfer) or

b) due to the energy dependence of τ.

In the near infrared region (0.1–0.3 eV) there is $\omega\tau \gg 1$ for most n-type semiconductors and therefore changes in the Faraday angle are only due to a changing mass whereas in the microwave region (≈ 1 GHz) relaxation time effects should be considered [107]. A theoretical treatment of the Faraday effect in the near infrared was given in Ref. [109, 110]. Alzamov [110] and Vorob'ev et al. [111] have pointed out that in addition to a change of the Faraday rotation in high electric fields, even in isotropic semiconductors like n-InSb, birefrigence will show up due to the anisotropy of the distribution function and due to nonparabolicity. In calculating the change of Faraday rotation $\vartheta = (2\pi d')(c n_r)^{-1} \operatorname{Re} \sigma_{xy} (\omega\tau \gg 1, \omega_c \ll \omega)$ in high fields, σ_{xy} was calculated be replacing the drifted MB for f_0 and by taking into account the nonparabolicity. Alzamov [110] has given the full high field conductivity tensor leading also to birefrigence. Early experiments on the Faraday rotation in n-InSb [108] in the spectral range of 0.14–0.21 eV showed a decrease of ϑ with increasing E attributed to the increasing effective mass. A refined analysis in the same material [111] is shown in Fig. 19. T_e was deduced from the change of ϑ with increasing E and from the change of the refractive index for light polarized $\parallel$ and $\perp$ to the applied field E. Figure 19 shows a "cooling effect" of T_e for small E predicted earlier by a calculation based on a drifted MB.

A detailed analysis on the hot electron Faraday effect in n-GaSb and n-Ge was made by Heinrich et al. [112, 113]. The principle objective of these measurements was not to deduce electron temperatures but to get information on the amount of carrier transfer from low lying conduction band valleys to the higher ones. The experimental arrangement is shown in Fig. 20. By using a sensitive analyzer a minimum resolution of 3×10^{-3} deg could be obtained. A He–Ne laser operating at 3.39 μm was used as a light source in the case of GaSb [112] and a 10.6 μm CO_2 laser in the case of Ge [113].

In both materials a decrease of the Faraday angle was found with increasing field due to the fact that the number of carriers in the Γ valley (GaSb) or L valleys (Ge) is reduced and the electrons are transferred to

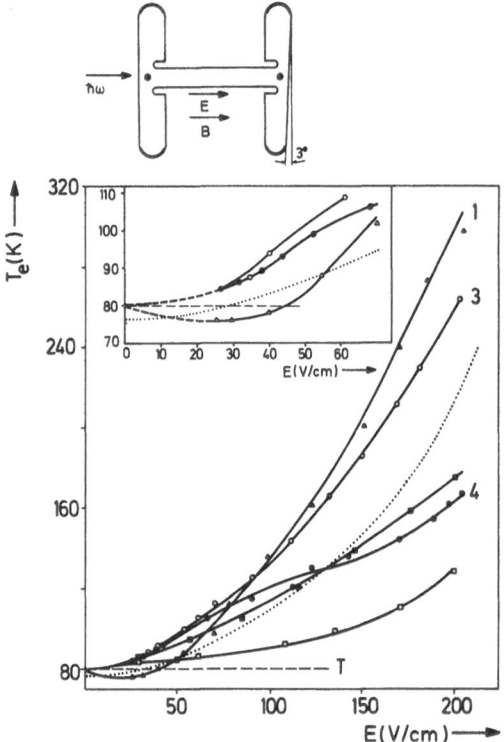

Fig. 19. $T_e(E)$ from Faraday effect and birefrigence measurements in n-InSb at 80 K. Upper half: sample configuration for Faraday effect. Lower half: $\triangle\ n = 3 \times 10^{14}\ cm^{-3}$ (1); $\bigcirc\ 1.2 \times 10^{15}\ cm^{-3}$ (3); $\bullet\ 4.9 \times 10^{15}\ cm^{-3}$ (4); 1, 3, 4: from Faraday effect measurements. $\blacksquare\ 6.5 \times 10^{14}\ cm^{-3}$; $\square\ 4.7 \times 10^{15}\ cm^{-3}$ both from birefrigence $\cdots\cdots$ calculated; from Ref. [111]

the higher L or X valleys, respectively. There they have a higher mass and do not contribute significantly to the rotation[14]. Use was made of the formulas given by Ipatova et al. for the Faraday angle in a many valley semiconductor having ellipsoidal minima [114].

In the case of n-GaSb a calculation based on an electron temperature model gave good agreement with the observed change of the Faraday angle of 0.15° at approximately 1 kV/cm where more than 10% of the electrons of the Γ valley were transferred to the L valleys (Fig. 21). It was shown that nonparabolicity of the Γ valley would only account for approximately 10% of the observed change in Faraday angle. The increase of the electron temperature turned out to be 30 K above the

[14] The effective masses for the Γ and the L valleys in GaSb are:

$$m_\Gamma = 0.047\,m_0,\ m_{L,t} = 0.143\,m_0,\ m_{L,l} = 1.23\,m_0\ (\text{Ref. [112]}).$$

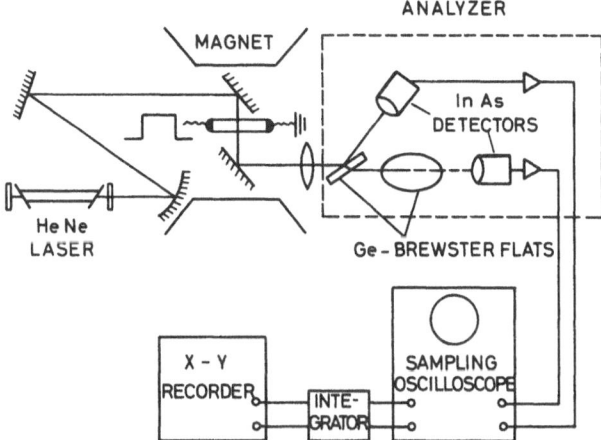

Fig. 20. Experimental arrangement for Faraday measurements. Brewster mirror plates in the analyzer are rotated 90° to each other. For measurements in n-Ge a CO_2 laser and $Hg_{1-x}Cd_xTe$ detectors are used ([112, 113])

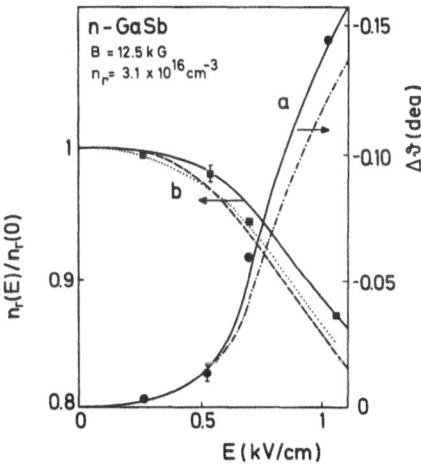

Fig. 21. (a) Decrease of Faraday angle with field (●); (—·—·—): reduction of multiple reflections; (b) decrease of carrier concentration in the Γ valley with field (■). (– – –): calculated using an electron temperature model (Ref. [102, 112]). (·····): results of a Monte Carlo calculation (Ref. [196])

lattice temperature of 300 K, at 1.1 kV/cm for a sample having a total carrier concentration of 6.8×10^{16} cm^{-3}.

In n-Ge the change of Faraday angle with electric field up to 5.5 kV/cm was investigated at $T = 200$ and 300 K. It was shown that both non-parabolicity in the L valleys and intervalley transfer to the X valleys

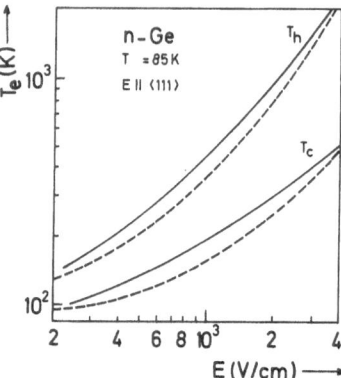

Fig. 22. Electron temperatures deduced from birefrigence measurements in n-Ge with $n = 5 \times 10^{14}$ cm^{-3}. (——) experimental values of T_e in $\langle \bar{1}11 \rangle$ valleys (T_h), and in [111] valley (T_c). (– – –) T_e deduced from balance equations. (After [116].) Above about 2 kV/cm carrier transfer to the $\langle 100 \rangle$ valleys is remarkable which is not considered in this figure

0.18 eV above the lower minima must be considered for a sufficient explanation of the observed decrease of the Faraday angle with electric field.

Another optical method for the investigation of hot electrons, the optical birefrigence was originally proposed by Schmidt-Tiedemann [6, 115] and later experimentally used for a study of the equivalent intervalley scattering in n-Ge between the $\langle 111 \rangle$ minima by Vorob'ev and coworkers [108, 116] in the near infrared (10.6 μm). Whereas in the absence of an electric field Ge being a cubic crystal does not show birefrigence, electric fields applied in a [111] direction cause repopulation of the electrons among the $\langle 111 \rangle$ valleys and the crystal becomes optically anisotropic, the [111] direction being the optical axis. The change of refractive index $\Delta n_r = n_{r,\parallel} - n_{r,\perp}$ for light polarized parallel and perpendicular to the [111] axis [108] is given by

$$\Delta n_r = \frac{2\pi e^2 n_{\langle 111 \rangle}}{n_r m_l \omega^2} \frac{(n_c/n_h) - 1}{3 + (n_c/n_h)} (m_l/m_t - 1)$$

where $n_{\langle 111 \rangle}$ is the total concentration in the $\langle 111 \rangle$ valleys and n_c and n_h are the concentrations in the [111] and in the $\langle \bar{1}11 \rangle$ valleys respectively. Changes of the refractive index for light polarized parallel and perpendicular to E between 1×10^{-5} and 2×10^{-4} could be detected in fields ranging from 0.1 to 4 kV/cm.

The population ratio of the cold to the hot valleys was deduced and temperatures in the hot and cold valleys were calculated from this data and from the balance equations using a MB distribution (Fig. 22). The

Table 2 (1). Determination of electron temperatures

Method	Material	Carrier conc. (cm^{-3})	Mobility (cm^2/Vs)	T (K)	E (V/cm)	B (kG)	T_e (K)	Ref.
$\mu(E)$	n-InSb	1×10^{14}	5.6×10^5 (20 K)	1.6 1.8 4.2	0.01–10		1.6 –50	[25]
	n-InSb	4×10^{14}	2.9×10^5 (20 K)	1.7 4.2	0.01–10		1.7–50	[25]
	n-InSb	1×10^{14}	3×10^4 (4.2 K) 1.9×10^5 (20 K)	4.2– 30	0.02–10		4.2–30	[28]
	n-InSb	1.19×10^{14}	7.5×10^4 (4.2 K)	4.2	0 – 0.17		4.2 –15	[26]
	n-InSb	1.25×10^{14}	6×10^4 (4.2 K)	1.35	0 – 0.2		1.35–17	[27]
	n-InSb	3.2×10^{14}	2.1×10^5 (10 K)	10	0.01– 0.4		10 –24	[38]
	n-InAs	2.5×10^{16}	2.5×10^4 (4.2 K)	4.2	0.25– 4		10 –28	[45]
$R_B(E)$, $\varrho_{xx}(E)$	n-InSb	1×10^{14}	5.6×10^5 (20 K)	1.6 1.8 4.2	0.01–10	0.05–11.2	1.6–70	[25]
	n-InSb	4×10^{14}	2.9×10^5 (20 K)	1.6 1.7 4.2	0.01–10	0.05–11.2	1.6–70	[25]
$R_B(E)$	n-InSb	2.2×10^{14} 4.4×10^{14}	2.1×10^5 (10 K)	1.5 20	0 – 0.3	0.16	1.5–20	[32]
	n-GaAs	3×10^{15}	1.2×10^3 (1.2 K)	1.22 12 27	0.1–40		Direct solution of BE for acoustic scattering	[53] [54]
g-factor	n-InSb	1.4×10^{14}	7.5×10^4 (1.7 K)	1.7	0.1–2.5	0.5	2.7–32	[62]
	n-InSb	1×10^{14}		1.6		1.5	1.6– 7.5	[63]
SdH	n-InSb	1.7×10^{15}	1.1×10^5 (1.3 K)	1.3	0– 0.072	0– 5	1.3– 5	[68]
	n-InSb	5.9×10^{15}	9.5×10^4 (4.2 K)	4.2	0– 0.140	0– 5.5	4.2–10.5	[76]
	n-InSb	1×10^{16}	6.6×10^4 (4.2 K)	4.2	0– 0.250	0– 7.5	4.2–16.5	[76]
	n-InSb	6.9×10^{16}	7.5×10^4 (4.2 K)	4.2	0– 1.8	0–20	4.2–26	[76]

Table 2 (continued)

Method	Material	Carrier conc. (cm⁻³)	Mobility (cm²/Vs)	T (K)	E (V/cm)	B (kG)	T_e (K)	Ref.
	n-InAs	1.2×10^{16}		1.7		0–13	1.7–12	[69]
	n-InAs	4.8×10^{16}	3.6×10^4 (4.2 K)	4.2	0– 1.0	0–45		[70]
	n-InAs	2.5×10^{16}	2.5×10^6 (4.2 K)	4.2	0– 0.3	0–12	4.2–12.2	[74]
	n-GaSb	1×10^{18}	5.0×10^3 (4.2 K)	4.2	0–15	30–50	4.2–18	[75]
	n-GaSb	3.4×10^{17}	4.3×10^3 (4.2 K)	4.2	0– 8	30–50	4.2–13.5	[75]
Noise	n-Ge	2×10^{14}		300	0–1800		300– 540 ($T_{n\perp}$)	[82]
	n-Ge	2×10^{14}		77	0–1600		77–3600 ($T_{n\perp}$)	[83]
	n-Ge	1.5×10^{14}		300	0–4000		300–1050 ($T_{n\parallel}$)	[87] [9]
	p-Ge	1.5×10^{14}		300	0–3700		300– 550 ($T_{n\parallel}$)	[87] [9]
	p-Ge	1.5×10^{14}		80	0–2000		80– 350 ($T_{n\perp}$) 80– 580 ($T_{n\parallel}$)	[87] [9]
Burstein shift	n-GaSb	3.5×10^{17}	7×10^3 (77 K)	77	0–150		77–100	[102]
Faraday effect	n-InSb	1.3×10^{15}	1×10^5 (90 K)	90	0– 250	2.5	90–170	[108]
	n-GaSb	6.8×10^{16}	3.1×10^3 (300 K)	300	0–1100	12.5	300–330	[112]
	n-InSb	3×10^{14}	5.6×10^5 (77 K)	80	0– 200	1.2	80–75–300	[111]
	n-InSb	1.2×10^{15}	2.6×10^5 (77 K)	80	0– 200	1.2	80– 260	[111]
	n-InSb	4.9×10^{15}	1.5×10^5 (77 K)	80	0– 200	1.2	80– 170	[111]
	n-InSb	1.7×10^{16}	1×10^5 (77 K)	80	0– 120	1	80– 100	[111]
Birefrigence	n-Ge	5×10^{14}		85	200–4000		100– 450 (T_c) 130–2000 (T_h)	[116]
	n-Ge	4.7×10^{15}		85	200–4000		90– 500 (T_c) 100–2000 (T_h)	[116]
	n-InSb	6.5×10^{14}	4.4×10^5 (77 K)	80	0– 200		80– 180	[111]
	n-InSb	4.7×10^{15}	1.8×10^5 (77 K)	80	0– 200		80– 130	[111]

Values for ohmic mobilities are given for lattice temperatures in parenthesis.

method is very sensitive and is capable of detecting changes in the carrier concentration in the hot and cold valleys as small as 1×10^{13} cm^{-13}. To our knowledge no application to other many valley semiconductors has been published.

At the end of the chapter in Table 2 the various investigations described in Sections 2.2.1–2.2.9 are summarized. The methods described in 2.2.1–2.2.4 are essentially restricted to low temperatures and dominating ionized impurity scattering for the momentum relaxation. Changes in T_e of the order of 1 K can be detected.

The methods described in Sections 2.2.5–2.2.9 are suitable to detect changes of T_e in much higher fields and in principle no restriction on the range of lattice temperatures exists.

3. Determination of Hot Electron Distribution Functions

3.1. Theoretical Foundation

If the calculation of transport properties is based on the Boltzmann equation, the main problem is the determination of the distribution function. Usually physical properties represent some mean value of a certain observable averaged over the distribution function. There are however some optical experiments which directly allow a probing of the electron distribution, and establish the dependence of the occupation number on the energy.

In Chapter 2 the high field distribution functions were characterized by a parameter T_e. If the scattering mechanisms and the range of electric fields are such, that the distribution function is characterized by $f_0 + f_1 \cos\theta$, analytical solutions can be given under certain conditions. The BE is then determined by a system of two equations if f is inserted into the drift and collision term. For acoustic phonon scattering, only analytical expressions are found if the equipartition approximation $(\hbar\omega_q \ll k_B T)$ is used (see [10]). Also solutions for acoustic and optical mode interactions are possible, if the collision term for the optical modes can be expanded in powers of $\hbar\omega_0/\varepsilon$. Reik and Risken [117] have succeeded in solving the BE for these scattering mechanisms in n-Ge taking into account the many valley band structure (see also Ref. [5]). The result was a MB like distribution with T_e generally different for each valley. Reik and Risken [117] have also been able to show that despite the expansion to terms quadratic in $\hbar\omega_0/\varepsilon$ the results are valid for average electron energies $k_B T_e \approx 2\hbar\omega_0$. The effect of equivalent intervalley scattering was also taken into account and an analytical expression can be given if the ratio of the intervalley (D_{ij}) to the optical coupling constant (D) is small. In this approximation the distribution function in valley j is given by [5, 10]

$$f_0^{(j)} = A^{(j)} \exp(-\varepsilon/k_B T_e^{(j)}) + \bar{\gamma} \sum_i c^{(i)}(\varepsilon) \exp(-\varepsilon/k_B T_e^{(i)})$$

where $\bar{\gamma}$ depends on the optical and intervalley coupling constants and the probabilities for emission and absorption of intervalley phonons[15]. $A^{(j)}$ is a constant depending on the particular valley. The summation extends over the number of equivalent valleys i. Thus the distribution function in valley j is given by a zeroth order MB distribution plus a linear combination of MB distributions from all valleys. For details we refer to Ref. [5] and [10].

In Ref. [8] the different methods for a solution of the BE in the diffusion approximation are reviewed with particular emphasis on the rate of electron-electron scattering.

Despite of the success of these analytical methods in explaining many hot electron phenomena they are unfortunately restricted to materials and ranges of electric fields for which no experimental determination of the distribution function has been possible.

In 1966 two methods were presented to calculate numerically distribution functions without introducing simplifying approximations. One of these methods is a Monte Carlo calculation introduced by Kurosawa [118] to the study of the properties of hot holes in p-Ge and the second method is an iterative method [119] also first applied to p-Ge. Since then considerable work has been devoted to Monte Carlo calculations as well as to the iterative method [11]. The calculations were mainly performed to deduce $v_d - E$ characteristics in III–V compounds like GaAs, InSb, InP where beside strong polar optical scattering, transfer to higher conduction band minima at high electric fields has to be considered. Both mechanisms make calculations based on the electron temperature concept or on a series of truncated spherical harmonics of the distribution function unsuitable. Although distribution functions have been calculated for a variety of materials and for a wide range of carrier concentrations in these III–V compounds, there are until now only three experimental investigations which may be compared with these calculations (Sections 3.2.3–3.2.5).

The main reason for the numerical calculations for n-type III–V compounds was the necessity to get a better understanding of the intervalley transfer process. This process is not only of interest for its own but has a great importance for possible device applications based on the Gunn effect [11].

---

[15] $\gamma = (D_{ij}/D)^2 (2N_q(\hbar\omega_{ij}) + 1)/\hbar\omega_{ij}$.

In the Sections 3.1.1 and 3.1.2 we shall briefly discuss the Monte Carlo method and the iterative method also known as path variable method. For a detailed description of these techniques we refer to Ref. [120] and [11].

3.1.1. Monte Carlo Method

In a Monte Carlo calculation, the path of a single electron in k-space is simulated using a computer. The electron is subjected to the influence of the electric field and of the various scattering mechanisms. In order to describe the motion of the electron the time which the electron drifts freely, the type of scattering process and the final state after the scattering are regarded to be random quantities determined by probability distributions. In this path of a single electron collisions with other electrons cannot be included [11].

Between two successive collisions the electron travels with constant velocity in k-space.

$$\mathrm{d}(\hbar k)/\mathrm{d}t = eE \qquad k = k_0 + eE(t-t_0)/\hbar. \tag{43}$$

A collision causes an instantaneous change of k which is determined by a probability distribution which depends on the total scattering rate $\lambda(k)$ given by

$$\lambda(k) = \sum_i \int W_i(k, k') \, \mathrm{d}^3 k'. \tag{44}$$

Following Fawcett et al. [11, 120] the probability distribution of the time of free flight can be derived in the following way: the electron starts for $t_0 = 0$ at k_0; if $P_d(k_0, t)$ is the probability that the electron drifts for a time t and is then scattered between t and $t + \mathrm{d}t$ then

$$P_d(k_0, t) = B_d(k_0, t) \, \lambda(k(t))$$

where B_d is the probability that the electron drifts for a time t without being scattered and $\lambda(k(t))$ is the probability per time that the electron will be scattered at k.

The arrival probability $B_d(k_0, t)$ is related to $\lambda((k(t))$ by the expression

$$B_d(k_0, t + \mathrm{d}t) = B_d(k_0, t) \, \{1 - \lambda(k(t)) \, \mathrm{d}t\}$$

where $(1 - \lambda(k(t)) \, \mathrm{d}t)$ is just the probability for the electron of not being scattered in the time interval $\mathrm{d}t$. Integrating the above equation yields

$$B_d(k_0, t) = \exp\left\{-\int_0^t \lambda(k(t')) \, \mathrm{d}t'\right\}$$

with $B_d(k_0, 0) = 1$ and $B_d(k_0, t \to \infty) \to 0$. Thus

$$P_d(k_0, t) = \lambda(k(t)) \exp\left\{ - \int_0^t \lambda(k(t')) \, dt' \right\}. \tag{45}$$

Random numbers with this probability distribution simulate the collision times. Since there are different scattering mechanisms possible, it is necessary to select one of them to terminate the free flight. This is done by generating random numbers which choose one of the $i = 1 \ldots n$ scattering mechanisms having the scattering rates $\lambda_i(k)$ according to the probability distribution [22, 122]

$$P_{di} = \lambda_i(k) \bigg/ \sum_{j=1}^{n} \lambda_j(k).$$

Further random numbers are required which determine the position of the electron after the scattering process. Since the magnitude of the wave vector after scattering is determined by energy conservation random numbers which simulate the scattering angle are generated according to $W(k, k') = W\{k, k', \angle(k, k')\}$ [22].

For the calculation, the k-space has to be devided in cells. In order to determine the distribution function, the time which the electron spends in each cell during the course of free flights is determined and yields the value of the distribution function for the particular cell. The transitions between the states due to the various scattering mechanisms are supposed to be instantaneous. In determining the distribution function in this way use is made of the "ergodic hypothesis"[16]. To obtain statistical convergence usually some ten thousand scattering processes have to be simulated in order to obtain a stationary distribution function. Macroscopic observables like the mobility and the mean energy are then easily calculated.

Fawcett et al. [120] have shown that their Monte Carlo procedure leads to a distribution function which satisfies the Boltzmann equation.

Monte Carlo techniques have also been used to calculate high field transport properties in magnetic fields [123] and have been applied to avalanche processes [124]. Preliminary attempts to include electron-electron scattering have also been made by using together with the Monte Carlo calculation an iterative method [125].

In Figs. 23 and 24 examples of calculated distribution functions and expansions of these functions in Legendre polynomials for n-GaAs and n-InSb are presented. In n-GaAs for fields well above Gunn threshold ($\approx 3\,\text{kV/cm}$) the distribution function (Fig. 23a) as well as its expansion

---

[16] For a discussion see e.g.: Becker, R.: Theorie der Wärme. Berlin-Heidelberg-New York: Springer Verlag, 1966, p. 99.

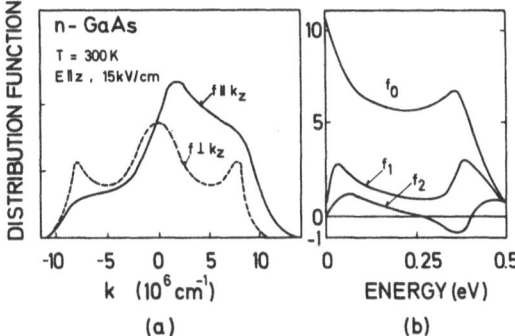

Fig. 23. Hot carrier distribution function for n-GaAs at 15 kV/cm parallel and perpendicular to an applied field; righthand side: expansion in Legendre polynomials. (After Ref. [11])

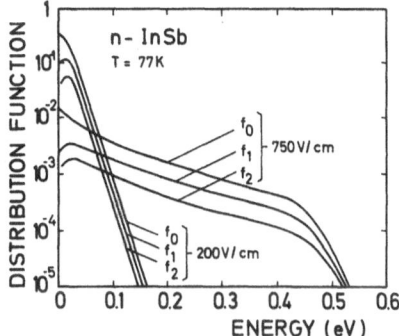

Fig. 24. Hot carrier distribution function in n-InSb for two fields in an expansion in Legendre polynomials. (After Ref. [11])

in spherical harmonics (23b) shows a considerable distortion caused by relatively weak polar optical scattering below 0.36 eV (the separation between the Γ and X valleys) and strong intervalley scattering above 0.36 eV. The second scattering process causes the negative f_2 term. Population inversion in f_0 below 0.36 eV is observable for fields in excess of 10 kV/cm. In n-InSb at 77 K (Fig. 24) the expansion of the distribution function shows that the first three coefficients f_n are comparable even at high fields.

3.1.2. The Iterative Method

In contrast to the Monte Carlo method, the iterative method is a procedure in which it is tried to solve the BE and to get directly a distribution function.

We follow the treatment given in Ref. [122]. The principle of the method is a reduction of the integro-differential BE to a differential equation of first order. If $f_n(k)$ is the n-th approximation of the distribution function, f_{n+1} will be calculated according to the following procedure: the change of the distribution function at k due to carriers scattered to this point in k-space is given by

$$g_n(k) = \int f_n(k')\, W(k', k)\, d^3 k' \,. \tag{48}$$

$g_n(k)$ can be calculated by integration (at least numerically). This expression is used to replace the second term of the collision term in the BE $(E \parallel z)$

$$\partial f/\partial t + (eE/\hbar)\,\partial f/\partial k_z + \lambda(k)\,f = g_n(k)\,. \tag{49}$$

Thus the integro-differential equation is reduced to a partial differential equation of first order which is solved by the introduction of new variables

$$k_z - (eE/\hbar)\,t = K_z \qquad t = \tau'$$

where the first of these variables is a constant for the collision free motion, therefore also called path variable and the procedure path variable method [119]. The partial differential equation is reduced to an ordinary differential equation. The solution is given by [122]

$$
\begin{aligned}
f(k, t) = f(k - eEt/\hbar, 0) \exp\left\{ - \int_0^t \lambda(k - eEt'/\hbar)\, dt' \right\} \\
+ \int_0^t dt'\, g_n(k - eEt'/\hbar) \times \exp\left\{ - \int_0^{t'} \lambda(k - eEt''/\hbar)\, dt'' \right\}
\end{aligned}
\tag{50}
$$

for $t \to \infty$ the solution yields $f_{n+1}(k)$. For $n \to \infty$ the iteration will be convergent but usually it is sufficient to consider only a few steps.

In practice, an additional self scattering process is introduced with a rate given by [11]

$$W_s(k, k') = W(k)\,\delta(k - k')\,.$$

Since the δ-function does not allow a momentum change to occur this process is not physically significant and $W(k)$ is an arbitrary function.

The addition of this process replaces $\lambda(k)$ by $\lambda(k) + W(k)$. $W(k)$ was taken to be $\Phi - \lambda(k)$ where Φ is a constant and has to be larger than the maximum of λ. Thus

$$f_{n+1}(k) = \int_0^\infty dt'\, g_n(k - eEt'/\hbar) \exp(-\Phi t')\,. \tag{51}$$

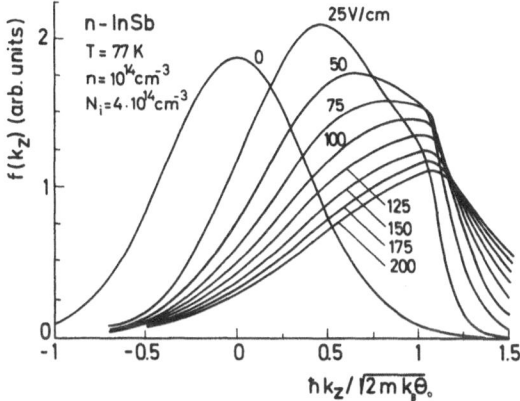

Fig. 25. High field distribution function in n-InSb for $E \parallel z$. For abscissa values above ≈ 1, optical phonon scattering causes the steep decrease of the functions (Ref. [122])

Rees [121] has shown that with the incorporation of self scattering, each iterative loop will be equivalent to a time step $1/\Phi$. If Φ is large enough, the resolution will be accurate enough to study the time development of the distribution function. Rees has also included the effect of electron-electron scattering which leads in the iteration to a dependence of the scattering rate on the distribution function in each step [121].

As an example, in Fig. 25 the distribution $f(k_z)$ for fields $E_z = 0$–200 V/cm is shown as a function of k_z for n-InSb at 77 K. The decrease of the distribution function for abscissa values larger than about 1 is caused by polar optical phonon emission.

3.2. Experimental Methods

3.2.1. Modulation of Intervalence Band Absorption by an Electric Field

The modulation of intervalence band absorption of p-Ge due to the application of high electric fields has been investigated by several authors [126–136]. These measurements have been the first attempt to deduce from the field induced change of the optical absorption direct evidence for the deviation of hot carrier distribution functions from the simple MB distribution.

The valence band structure of Ge near $k=0$ consists of three doubly degenerate bands, two of these are degenerate at $k=0$ and the third is split off by 0.295 eV by spin orbit interaction [136–140]. The valence band structure was deduced from optical absorption measurements in the energy range from 0.15–0.65 eV [137, 138] where two peaks and a

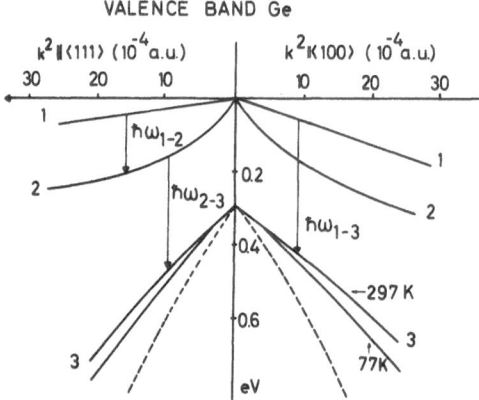

Fig. 26. Valence band of Ge. (———) Ref. [136], (– – –) Ref. [140], indicated transitions: 1–3, 2–3, 1–2; a.u. represents atomic units

shoulder were found. Calculations based on second order $k \cdot p$ perturbation theory were performed by Kahn and Kane [139, 140]. According to Kane, Band 1, the heavy hole band is approximately parabolic, in Band 2, the light hole band, the effective mass increases with energy and in Band 3, the splitt-off band it decreases with energy. Due to the different curvature of Bands 2 and 3 the absorption spectra caused by direct 1–3 and 2–3 transitions do not overlap. It was realized early that in the Kane model the non parabolicity of the split-off band was over-estimated. Later Fawcett [141] refined the Kane theory by taking into account $k \cdot p$ interactions between the valence band and two conduction bands exactly and by considering the next (third) nearest conduction band as a perturbation. Using this model Arthur et al. [142] have been able to calculate the detailed structure induced into the absorption spectrum by a small increment of lattice temperature. The comparison with experiment showed that calculations based on the Kane model deviated more pronounced from the data than the Fawcett model. The main difference between these two models concerns the shape of the Band 3: in Fawcett's calculation it is much more parabolic. Christensen [136] has shown that a $k \cdot p$ calculation based on a four-band model [143], taking into account only the lowest lying conduction band, yields almost the same results for the band structure than Fawcett's model as shown in Fig. 26. By taking into account the temperature dependence of the energy separation to the nearest conduction band with Γ_2^- symmetry a temperature dependence of the shape of the split-off band resulted. This effect was found to be in good agreement with experiments by Kessler and Kneser [144] on the temperature dependence of the non parabolicity of Band 3. The energetic separation of the split-off Band 3 from the top of the valence band is found to be independent of temperature. The temperature dependence of the curvature of the third band becomes however only noticeable for energies higher than approximately 0.4 eV.

Generally the absorption constant K at a photon energy $\hbar\omega$ due to a direct transition from the initial (i) to the final state (j) is given by [145]

$$K = \frac{8\pi^2 e^2 \hbar}{n_r m_0^2 c} \frac{1}{\hbar\omega} \frac{1}{8\pi^3} \int d^3 k |\langle j| \, a \cdot p \, |i\rangle|^2 \, \delta(\varepsilon_j(k) - \varepsilon_i(k) - \hbar\omega)$$
$$\cdot (f_i(k) - f_j(k)) \tag{52}$$

where a is the polarization vector of the radiation. For the $i-j$ (1–3) transitions, using as an approximation parabolic bands and considering that the transitions are forbidden, K is given by [146]

$$K \sim C(p_1/N_v)(\hbar\omega - \Delta)^{3/2} (\hbar\omega)^{-1} \exp\left[-\frac{m_{ij}}{m_i}\left(\frac{\hbar\omega - \Delta}{k_B T}\right)\right] \tag{53}$$

where C is a constant and $1/m_{ij} = 1/m_j - 1/m_i$. p_1 is the hole concentration in Band 1 and N_v the density of states factor. Band 3 is totally empty in the temperature range of interest.

Both formulas indicate that the absorption at a given energy will depend on the occupation probability in Band i. In order to deduce an unique distribution function in Band 1 from absorption measurements following assumptions have to be made: the warping of the bands has to be neglected, all bands are considered to be sharply defined, and no indirect transitions contribute to the observed absorption coefficient. Then an unique relation can be established between the energy of the absorbed photon and the energy of the hole [129]

$$\hbar\omega_{1\to3}(k_j) = \varepsilon_3(k_j) - \varepsilon_1(k_j). \tag{54}$$

The momentum of the photon can be neglected. For 2–3 transitions a similar relation holds.

If a one-to-one correspondence is established between a certain photon energy $\hbar\omega$ and the energy of the hole, the absorption constant yields direct information on the occupation probability according to Eq. (53). In thermal equilibrium the occupation probability is determined by MB statistics so that an energy dependent proportionality factor $C'(\hbar\omega)$ can be determined

$$K_{th}(\hbar\omega) = C'(\hbar\omega) f(\varepsilon_1). \tag{55}$$

If the band structure does not change in an applied electric field and if the transition matrix element is independent of the electric field, $C'(\hbar\omega)$

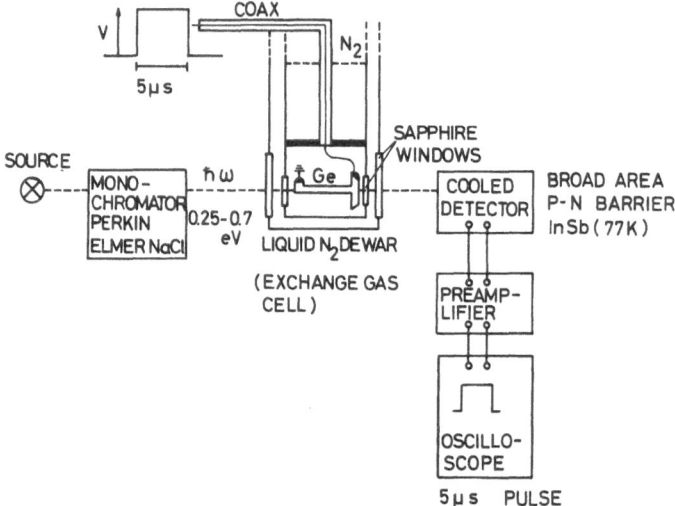

Fig. 27. Experimental arrangement for measuring change in intervalence band absorption due to high electric fields. (After Ref. [129])

will relate the absorption constant measured in an electric field K_E to the distribution function. Thus changes in the occupation probability of Band 1 and Band 2 can in principle be deduced.

The experimental arrangement used in the high field experiments is shown in Fig. 27. The electric field is applied in pulses of 2–5 μs duration. The light source is either a tungsten filament or a globar source. A monochromator is placed either between the light source and the sample or between the sample and the diffused InSb junction detector having a rise time of 1 μs. The sapphire windows of the cryostat shown in Fig. 27 were replaced by CaF_2 windows if polarized light was used. With a boxcar integrator instead of the oscilloscope a relative change in transmission of 4×10^{-4} could be measured. The spectral resolution was about 10 meV. Measurements were made of the ratio of change ΔI in transmitted intensity during the pulse to the transmitted intensity in zero field. From the change in transmitted intensity the change of the absorption constant is calculated by taking into account the reflectivity.

In the first published experiments Brown and Paige [126] reported the change ΔK between 0.25 and 0.65 eV at 93 K and 293 K in fields up to 1480 V/cm. At 93 K, a comparison of the field induced changes of K with changes due to an increase ΔT of the lattice temperature showed that up to 350 V/cm the high field distribution remains a MB distribution. At a lattice temperature of 293 K however the field induced changes were completely different from the ΔT effect.

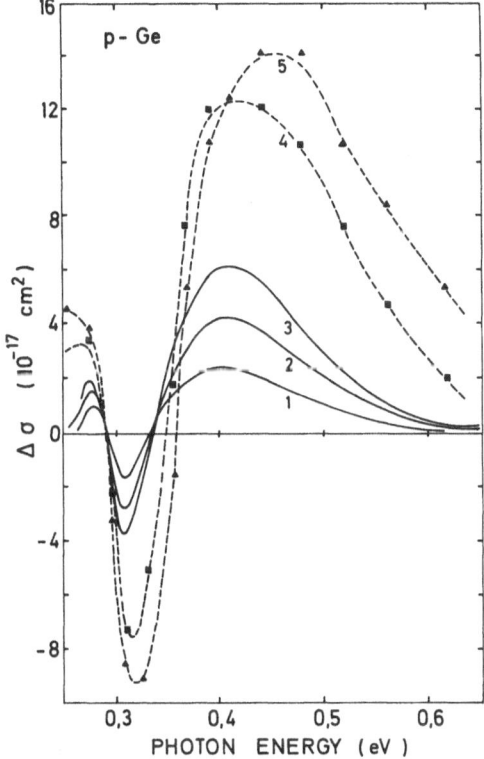

Fig. 28. Change of absorption cross-section in p-Ge. Curves 3, 2, 1: $p = 6 \times 10^{14}$, 3.9×10^{15}, 9.1×10^{15} cm^{-3}, $T = 93$ K, $E \approx 300$ V/cm ($\mu E^2 = 1.4 \times 10^9$ Vs^{-1}) (Ref. [127]); Curves 4, 5: $p = 1.9 \times 10^{15}$ cm^{-3}, $T = 77$ K, ■: $E = 810$ V/cm: ▲: $E = 1730$ V/cm (Ref. [129])

Figure 28 shows the change of absorption cross section $\Delta\sigma = \Delta K/p$ for various fields between $E = 300$ V/cm to 1730 V/cm. The data are taken from two different investigations [127, 129]. The spectrum above about 0.3 eV arises due to 1–3 transitions and below due to 2–3 transitions.

As far as the heavy hole transitions are concerned the following results can be deduced. At low energies of the heavy holes the electric field causes a depopulation and therefore a decrease of absorption. A positive modulation which is shifted to higher energies at higher fields indicates an increased occupation probability for energies of the heavy holes higher than approximately 0.02 eV. In the region of 2–3 transitions only positive modulation is observed.

The main difference between the work of Pinson and Bray [129] performed at 77 K and of Brown, Paige and Simcox [127] (Fig. 28) concerns the dependence of the absorption cross section on the carrier concentration. Whereas Pinson and Bray do not find an influence of p between 5.9×10^{14}–1.9×10^{15} cm^{-3} on the magnitude or shape of the absorption cross section, Brown et al. [127] already observe for a carrier concentration of 3.9×10^{15} cm^{-3} a marked decrease in $\Delta\sigma$ compared to 6×10^{14} cm^{-3} at a lattice temperature of 93 K at $E \leq 350$ V/ cm. The comparison between the different samples was made at equivalent power input from the field ($e\mu E^2 = $ const) to avoid misinterpretations due to different mobilities. Since according to [127] for a concentration of 9.1×10^{15} cm^{-3} the change in absorption cross section is again decreased, differences between the distribution functions in the samples are anticipated. It is suggested that the distortion of the MB distribution is most effective in the sample with the lowest carrier concentration. Somewhat in contrast to these experiments and in agreement with [128], Vasileva et al. [133] did not find substantial differences in $\Delta\sigma$ for samples with $p = 6.5 \times 10^{14}$ cm^{-3}, and 3×10^{15} cm^{-3} in fields up to 960 V/cm at a lattice temperature of 85 K if the comparison was made for the same $e\mu E^2$ values.

Pinson and Bray [129] have performed also measurements on the orientational anisotropy of the absorption coefficient at 77 K. They have determined $\Delta\sigma$ for $E \parallel [111]$ and $E \parallel [100]$ directions, the radiation propagating parallel to E. $\Delta\sigma$ is somewhat higher for $E \parallel [100]$ compared to $E \parallel [111]$ (at 1730 V/cm $\Delta\sigma_{111} = 1.4 \times 10^{-16}$ cm^2 and $\Delta\sigma_{100} = 1.8 \times 10^{-16}$ cm^2 for a photon energy of 0.45 eV). The orientational anisotropy in K is observed at the same fields where it is noticeable in the $j - E$ characteristics. The dependence of K on the direction of E with respect to the crystallographic axes may be caused by a hot carrier effect.

Carrier distribution functions were only deduced by Pinson and Bray [129] from measurements in undefined polarized light. Brown et al. [127] indicated only that the distribution functions at about 350 V/cm at $T = 93$ K already deviated from the MB as evidenced by a comparison with absorption data from a ΔT experiment. The distortion was more remarkable at higher lattice temperatures where again differences in $\Delta\sigma$ between samples of carrier concentration ranging from 6×10^{14}–9.1 $\times 10^{15}$ cm^{-3} were observed. It was stated that the differences should be attributed in part to the influence of hole-hole scattering. Measurements with polarized light [131–133, 147] were later all analysed with respect to the deduction of distribution functions (see Section 3.2.2).

From the data presented in Fig. 28 only the heavy hole distribution can be deduced. The split-off Band 3 is broadened to an extent of 0.02 eV due to the short life time of holes in this band. McLean and Paige [148] have explained this broadening by high energy phonon scattering into Band 1. This broadening has more effect on 2–3 transitions since these

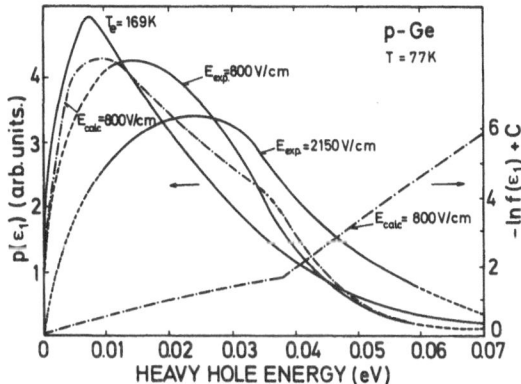

Fig. 29. $p(\varepsilon_1)$ (distribution of heavy holes) vs energy in p-Ge for 800 and 2150 V/cm (Ref. [129]), $p = 1.19 \times 10^{15}$ cm^{-3}, $E \parallel [100]$. Curve for 800 V/cm is compared with a MB distribution of equal mean energy ($T = 169$ K) and with a calculated curve by Budd [119] (—·—·—). Logarithm of probability distribution shows sharp kink at the optical phonon energy 0.037 eV

occur in a narrow range of k. However some uncertainty will also be introduced to 1–3 transitions in the vicinity of $k = 0$.

In analysing their results (Fig. 28) Pinson and Bray [129] have restricted the determination of $f(\varepsilon_1)$, the distribution function of heavy holes, to a range of heavy hole energies from 0.015–0.06 eV. The upper limit is caused by the fact that for high energies the intervalence band absorption becomes weak and intrinsic absorption dominant. Outside this range the distribution functions were obtained by extrapolation and checked by the condition that the total hole concentration remained constant. The experimental distribution functions are therefore most reliable where only a few carriers are below 0.015 eV and above 0.06 eV. This was the case for intermediate fields around 800 V/cm.

Since the exact shape of Band 3 was not known when Pinson and Bray [129] made their experiments, a careful experimental investigation of the temperature dependence of K was made. However in deducing the band shape of Band 3 its non-parabolicity above 0.45 eV was overestimated as was later pointed out by Kessler and Kneser [149].

The distribution functions of the heavy holes for $E = 800$ and 2150 V/cm are shown in Fig. 29. In this figure $p(\varepsilon_1)$, proportional to the number of carriers having a certain energy is plotted vs heavy hole energy. The distribution at 800 V/cm is compared with a MB distribution of the same mean energy that corresponds to $T_e = 169$ K. The dash dotted curve is taken from Budd [119] based on a path variable calculation. The calculated occupation probability distribution also shown in this figure has a sharp kink at approximately the optical phonon energy of 0.037 eV.

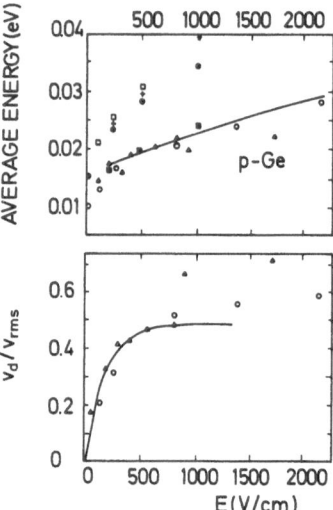

Fig. 30. Upper part: average energy vs electric field in p-Ge: experimental data: ○, Ref. [129]; △: Ref. [133]; ■: Ref. [136]; also included data on influence of uniaxial pressure of 5900 (●), 8850 (×), and 11 800 kp/cm^2 (□) on average energy Ref. [136]. ([129]: $T = 77$ K; [133, 136]: $T = 85$ K). Calculated data (——) Ref. [119], ▲ Ref. [118]. Lower part: increase of v_d/v_{rms} with electric field; symbols are the same as in upper part

This distribution is very well approximated by two Maxwellian distributions with different effective temperatures above and below the optical phonon temperature. Generally the experimental and calculated data show that a depletion of the distribution function occurs at energies higher than 0.037 eV whereas in an intermediate energy range the actual distribution function leads to a higher occupation probability than the MB distribution leading to the same mean energy.

For the experimentally determined distribution functions the increase of mean carrier energy with electric field is calculated by averaging over $f(\varepsilon_1)$. In Fig. 30, $\langle \varepsilon \rangle$ is shown vs E. Average energies deduced by different authors are compared. The full line represents a calculation by Budd [119] and the full triangles are based on a Monte Carlo calculation by Kurosawa [118]. In the lower part the drift velocity v_d, as found in $j-E$ characteristics related to the root mean square velocity (v_{rms}) is shown as a function of E. Beside experimental data, again calculated results from Budd and Kurosawa are included. The experimental data of Pinson and Bray [129] and of Christensen [136] on the average energy are in good agreement with each other. The data of Vasileva et al. [133] were taken with a sample of a carrier concentration of 6.5×10^{14} cm^{-3}. Her values for v_d/v_{rms} are higher than Pinson and Bray's data[17]. Generally the increase of v_d/v_{rms} above the calculated values will be influenced by the fact that the experimental drift velocities are total drift velocities due to heavy and light holes whereas the calculated ones are only due to heavy holes.

These two diagrams essentially describe the properties of heavy hot holes in p-Ge: the average energy of the carriers increases after an initial rise only slowly and does not achieve the optical phonon energy $\hbar\omega_0 = 0.037\,\text{eV}$ even at 2000 V/cm. The plot v_d/v_{rms} indicates that in p-Ge much of the kinetic energy is direct parallel to the electric field rather than at random. Thus in an expansion of the distribution function in Legendre polynomials not only the first two terms will be significant but higher order terms have to be included.

In order to explain the large v_d/v_{rms} values found in p-Ge, Pinson and Bray [129] have suggested as a qualitative model a cyclic streaming motion of the carriers: heavy holes are accelerated by the field up to the optical phonon energy; at this energy the holes suddenly emit an optical phonon and start again at low energy. Randomizing interaction processes like acoustic phonon scattering are weak in the range of temperatures, fields, and concentrations used in the experiments.

3.2.2. Determination of the Anisotropy of the Distribution Function by Electric Field Induced Dichroism

The hot carrier distribution function can exhibit a considerable anisotropy in momentum space if strong inelastic scattering mechanisms are important (see 3.1.1). For optical phonon scattering e.g. a large positive f_2 term will occur in an expansion of the distribution function in Legendre polynomials [Eq. (11)], representing the elongation of the distribution function in field direction. A method which can reveal this anisotropy is the dependence of the absorption of polarized light with respect to the orientation of the applied electric field E.

A determination of the anisotropy of the distribution function in high fields has until now only be performed in p-Ge [130–133, 147, 150, 151]. The absorption at a certain energy will depend on the angular distribution of holes on a constant energy surface and the angular form of the transition probability. According to Kane [130] the transition probability for 1–3 transitions in p-Ge is proportional to $\sin^2(a, k)$, where (a, k) is the angle between the polarization vector a and k. If the 4th and higher order terms in the expansion of the distribution function

---

[17] The discrepancy between Vasileva et al. results [133] and the results of [129, 136] may be caused by the fact that in [133] the non-parabolicity of Band 3 was overestimated.

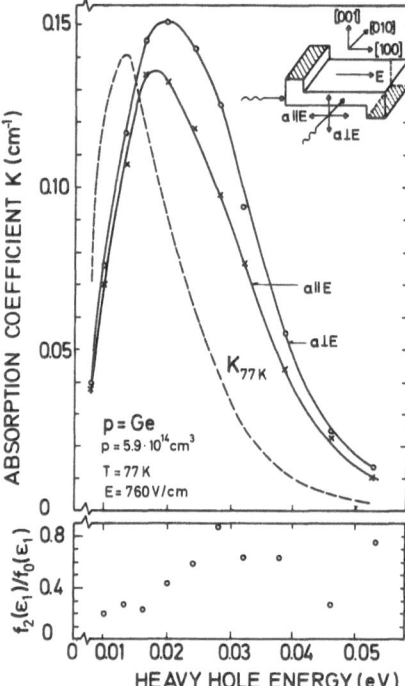

Fig. 31. Influence of high electric field on the polarization dependence of intervalence band absorption. Units of abscissa of K diagram are transformed in heavy hole energies. (After Ref. [130])

can be neglected, the anisotropy ratio $f_2(k)/f_0(k)$ can be determined from measurements in the configuration $a \parallel E$ and $a \perp E$ [130, 131]

$$\frac{f_2(k)}{f_0(k)} = \frac{10(1 - K_\parallel(\hbar\omega)/K_\perp(\hbar\omega))}{(2 + K_\parallel(\hbar\omega)/K_\perp(\hbar\omega))}. \tag{56}$$

In deriving this expression[18], again the assumption was made that the photon energy $\hbar\omega$ is uniquely related to the heavy hole energy, so that warping of the bands is neglected. The first investigations showed that in the range of the 1–3 transition $K_\perp > K_\parallel$ (Fig. 31) in agreement with the expectations[19].

---

[18] In Ref. [131] a different expression is given since a surface spherical harmonic expansion is used: $f(\varepsilon, E, \theta, \varphi) = \Sigma_{l,m} f_{l,m}(\varepsilon, E) Y_{l,m}(\theta, \varphi)$. In both investigations $E \parallel [100]$ ([130, 131]).
[19] Free carrier absorption, which would lead to a $\cos^2(a, k)$ dependence of the transition probability [see e.g. Fomin, N. V.: Sov. Phys. Solid State **2**, 566 (1960)] does not contribute significantly to the observed K in the range of photon energies 0.3–0.6 eV. At 0.35 eV K due to free carrier absorption is about 50 times smaller than K due to 1–3 intervalencebandabsorption [146] for the sample used.

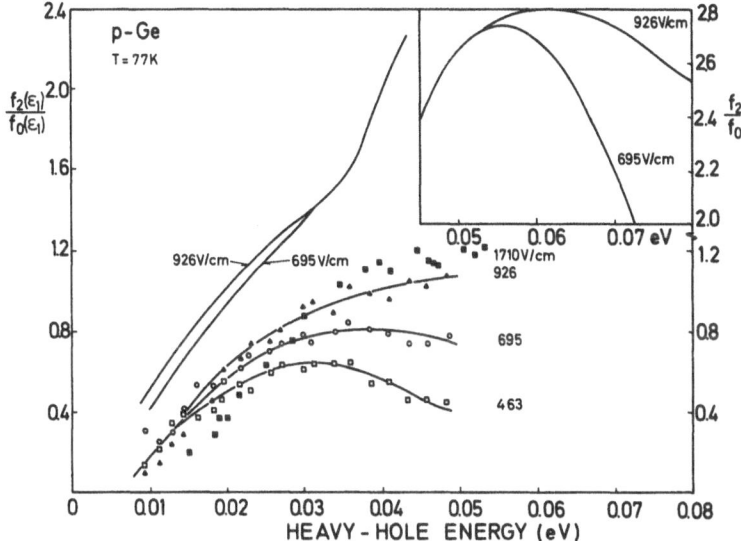

Fig. 32. $f_2(\varepsilon_1)/f_0(\varepsilon_1)$ vs heavy hole energy $\square$, $\bigcirc$, $\triangle$ exp. data from Ref. [132]; $\blacksquare$: Ref. [133]; (———): calculation Ref. [119]

In Fig. 32 the field dependence of f_2/f_0 is shown. The experimental results are from Ref. [132] and [133] and were obtained for samples with a carrier concentration of $2.77 \times 10^{15} \, cm^{-3}$ at 77 K and $6.5 \times 10^{14} \, cm^{-3}$ at 85 K respectively. The calculated values according to Budd [119] are too high. Budd states that carrier-carrier scattering, which was not included in his path variable calculation would cause a diminishing of the f_2/f_0 ratio. His results qualitatively describe the observed dependence of f_2/f_0 on E, the peak values occuring in the experiments at half the energy and are about three times smaller.

In the course of a study of the polarization dependent absorption in p-Ge, Baynham and Paige [147] and Vasileva et al. [133] have also got information on the light hole distribution $f(\varepsilon_2)$ in high fields. In order to establish the relationship between ε_2 and $\hbar\omega_{2-3}$ the transitions between the Bands 1 and 2 were investigated. The interpretation of the 2–3 transitions is complicated due to the life time broadening and due to a critical point in the joint density of states in this transition. From the relationship between ε_1 and $\hbar\omega_{1-3}$ and between ε_1 and $\hbar\omega_{1-2}$ the dependence of ε_2 on $\hbar\omega_{2-3}$ was deduced. In addition in Ref. [133] also data of the influence of high fields up to 1710 V/cm on the change of absorption cross section for the 1–2 transitions are given. For an electric field of 500 V/cm Baynham and Paige [147] got for the distribution of light holes approximately a MB distribution at 105 K for a lattice temperature of 77 K. Only $f_0(\varepsilon_2)$ could be determined.

Vasileva's [133] experimental data on the 1–2 transition were not conclusive enough to allow an unambiguous determination of $f(\varepsilon_2)$. There are several complications in the analysis of the data: strong interband scattering from the light hole to the heavy hole band, the dependence of the $\Delta\sigma_{1-2}$ curves on carrier concentration whereas no such dependence in the $\Delta\sigma_{1-3}$ curves was found. Only estimates of the effective temperature for the distribution of the light holes were made for fields of 960 V/cm and 1710 V/cm which yielded 500 and 700 K which are relatively high compared to 105 K at 500 V/cm from the analysis of the 2–3 transitions [147].

In a further attempt to get more information on the light hole distribution function Vorob'ev and Stafeev [150] have analysed the spontaneous long wave length infrared radiation (0.09–0.21 eV) originating from direct transitions of hot light holes into the heavy hole band. Radiation was observed for fields in excess of 1.5 kV/cm and has been analysed for fields up to 5.8 kV/cm. Its polarization with respect to the electric field was also observed so that $f_0(\varepsilon_2)$ and $f_2(\varepsilon_2)$ were deduced for an energy interval $\varepsilon_2 = 0.1$–0.25 eV. For an electric field of 5.8 kV/cm the ratio f_2/f_0 was found to be between 2 and 3. The authors have drawn the following conclusion from these experiments: no effective temperature can be assigned to the light hole distribution function and more terms in the expansion should be necessary to describe the distribution function properly. Nevertheless in a further paper [151] the authors have given temperature values of $T_e = 450$ K for $E = 4$ kV/cm and $T_e = 600$ K for $E = 5.2$ kV/cm corresponding to a MB distribution in the light hole band.

All the experiments discussed until now are based on the assumption that warping of the valence bands can be neglected. Christensen [136] has recently made a calculation based on the four band $\mathbf{k} \cdot \mathbf{p}$ model to check how close a relationship between a particular hole energy and a photon energy is established in reality. His results are shown in Fig. 33. For hot carrier distribution functions which have finite values between the energies indicated in Fig. 33 and are zero outside, the corresponding change of absorption cross section was calculated. There exists only an approximate correspondence between the range of photon energies which will lead to a 1–3 transition and the initial hole·energies in Band 1.

Christensen [136] has made measurements of the absorption cross section of p-Ge ($p = 2.1 \times 10^{15}$ cm^{-3}) at 85 K for fields up to 1000 V/cm. He has also considered the influence of an uniaxial stress $\|[111]$ up to 11 800 kp/cm^2 on the absorption spectra. As can be seen from the lower

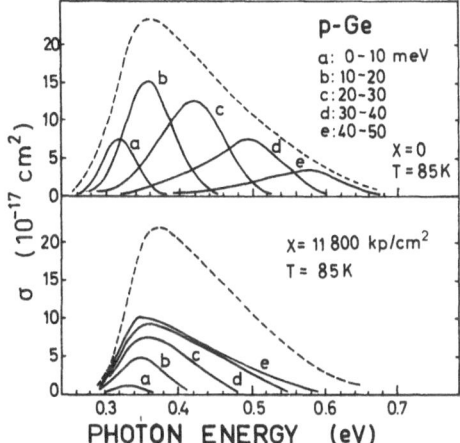

Fig. 33. Calculated absorption spectra for hot electron distribution functions which are zero outside the indicated range of heavy hole energies (a–e). The appreciable overlap of the curves is caused by warping. Uniaxial stress increases the overlap. (After Ref. [136])

part of Fig. 33 it is not at all possible to relate a photon energy to a hole energy in high uniaxial stress. In such a situation it is no longer possible to determine uniquely a distribution function. But the absorption spectrum can be calculated by using a parametrized distribution function and adjusting the parameters. The effect of strain on the intervalence band absorption had to be included in the matrix elements between the valence bands [136, 143]. In stressed p-Ge the degenerate Bands 1 and 2 are splitted and the absorption depends already in zero field on the direction of polarization of light with respect to the stress axis (Fig. 34).

Measurements were made for E and stress parallel to the [111] direction. In Fig. 35 the experimental data for $X = 0$ stress and for $X = 11\,800$ kp/cm² are presented for polarization $\parallel$ and $\perp$ to E. In Fig. 35 also data from Baynham and Paige [147] for a material with $p \approx 9 \times 10^{15}$ cm⁻³ and $T = 77$ K are included. For $X = 0$, over most of the spectral region the anisotropy is such that $\Delta\sigma_\perp > \Delta\sigma_\parallel$, and it increases in higher fields due to the increasing number of carriers in field direction. For $X \neq 0$, there is a competition between the effect of field and the effect of stress. Due to the stress, the carriers occupy states in k-space perpendicular to stress (Fig. 34) and $\Delta\sigma_\parallel > \Delta\sigma_\perp$. In high fields the carriers are forced into states parallel to the field and therefore a tendency for $\Delta\sigma_\perp > \Delta\sigma_\parallel$ exists. The calculated absorption cross section indicated by full and dashed lines in Fig. 35 are based on a MB distribution. Since earlier measurements had shown [131, 133, 147] that the distribution function can be approximated by two MB distributions with two

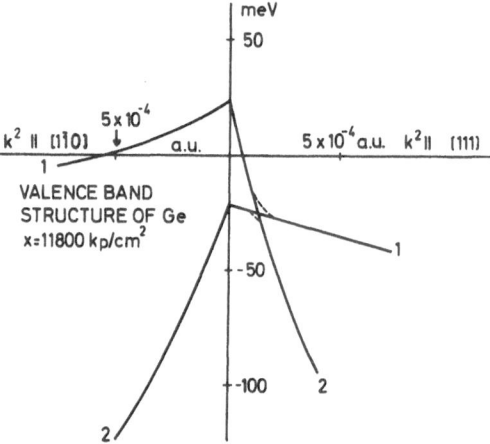

Fig. 34. Valence band structure of p-Ge for uniaxial stress X in [111] (Ref. [136])

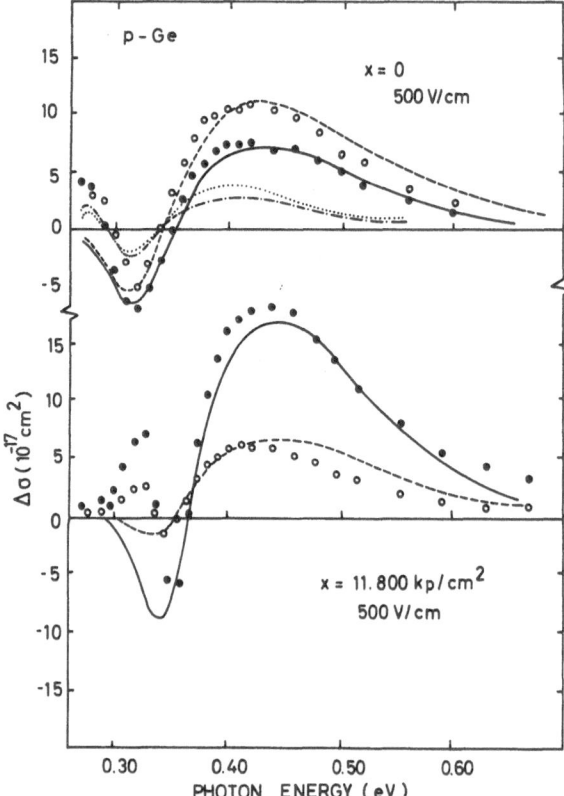

Fig. 35. Comparison of measured and calculated change of absorption cross section for $E = 500$ V/cm and $X = 0$ and $X = 11\,800$ kp/cm². Data from Ref. [136] (○, ●, $T = 85$ K, $p = 2.1 \times 10^{15}$ cm⁻³) and from Ref. [147] (— · — · —, ……, $T = 77$ K, $p \approx 9 \times 10^{15}$ cm⁻³) (———) and (– – –) are calculated data, Ref. [136] for $\Delta\sigma_\parallel$ and $\Delta\sigma_\perp$

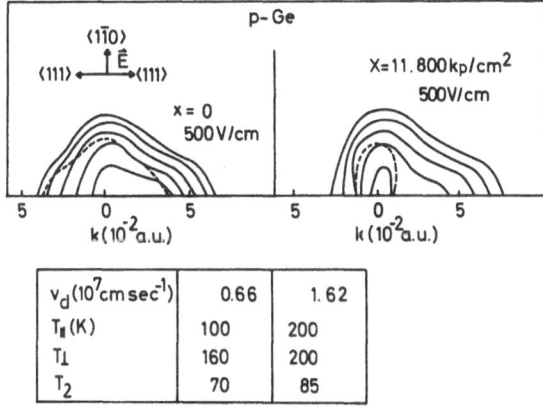

$v_d (10^7 \text{cm sec}^{-1})$	0.66	1.62
$T_\parallel$ (K)	100	200
$T_\perp$	160	200
T_2	70	85

Fig. 36. Curves of equal occupation probability in a k-plane for $X = 0$ and $X = 11\,800$ kp/cm². Occupation probability beginning from the center 1, 0.5, 10^{-1}, 10^{-2}, 10^{-3}, 10^{-4}. (– – –): optical phonon energy; Ref. [136]. Below parameters of the distribution function are given

effective temperatures [20], below and above the optical phonon energy $\hbar\omega_0$, Christensen [136] also assumed such a function. To reproduce the anisotropy of the spectra, the temperature T_1 for the energy region below the optical phonon energy was splitted in $T_\parallel$ and $T_\perp$ the "longitudinal" and "transverse" carrier temperature.

$$1/T_1(\gamma) = \cos^2\gamma/T_\parallel + \sin^2\gamma/T_\perp$$

where γ is the angle (k, E). T_2 the carrier temperature above $\hbar\omega_0$, was found by analyzing high field mobility data for the power loss per carrier. It was adjusted to make $ev_d E = P_{\text{opt}}$ the power loss to optical modes (see Section 4.1). So the distribution function was a function $f(T_\parallel, T_\perp, T_2, v_d, k)$. In Fig. 30 we have included Christensen's data on the change of mean energy with field for zero stress and stress up to $11\,800$ kp/cm² which are in good agreement with results of Ref. [129].

In Fig. 36 calculated curves of equal occupation probability are shown for $E = 500$ V/cm for zero stress and $11\,800$ kp/cm² (normalized to 1 at $k=0$ and then decrease to 0.5, 10^{-1}, 10^{-2}, 10^{-3}, 10^{-4}). The dashed line represents the optical phonon energy. In the lower part the parameters of the distribution functions which were used to calculate the upper curves are given. With these values the fit to the absorption data of Fig. 35 was performed. The main result of the applied stress is an increase of the anisotropy of the distribution function as demonstrated in Fig. 36.

---

[20] See Section 4.2, Ref. [208].

In summarizing the data presented in Sections 3.2.1 and 3.2.2 the following results were obtained:

a) the field dependence of intervalence band transitions yields information on the heavy and to some extent on the light hole distribution function,

b) measurements in polarized light reveal the anisotropy of the high field distribution function in k-space (dichroism)[21],

c) a unique distribution function is only obtained if warping of the bands is neglected,

d) by taking into account the real band structure only a fitting procedure yields a distribution function,

e) due to the strong interaction between hot heavy holes and optical phonons the occupation probability at energies higher than the optical phonon energy is reduced considerably,

f) up to fields of approximately 1000 V/cm a "two-temperature MB distribution" represents as a good approximation the observed distribution functions. The "temperature" for the distribution function above the optical phonon energy is always close to the lattice temperature.

Modulation of intervalence band absorption by high electric fields has only been used in p-Ge to determine distribution functions. It has contributed a lot to a better understanding of hot carrier properties in this material. Therefore it might be of interest to consider why this method has not been used in other semiconductors. No intervalence band transitions have been detected in p-Si since the split-off valence Band 3 is only 0.04 eV below Band 1 at $k = 0$ [146] so that it is unsuitable for such a method. Semiconductors with zincblende structure like the III–V compounds have reduced symmetry properties in comparison with Group IV elements. A k-linear term in the $\varepsilon(k)$ relationship for Band 1 removes the double degeneracy and the maxima are shifted away from $k = 0$. In InSb the spin orbit splitting Δ is about 0.82 eV such that it is much larger than $\Delta\varepsilon_g$ and the electronic fundamental absorption tends to mask intervalence band transitions. Another material which may be of interest for hot carrier experiments is p-GaAs. In GaAs, with $\Delta = 0.33$ eV and $\Delta\varepsilon_g = 1.53$ eV, the intervalence band transitions do not coincide with the electronic fundamental absorption. Absorption coefficients have been determined [146] for photon energies between 0.2 and 0.8 eV and show $1 \rightarrow 3$, $1 \rightarrow 2$, and $2 \rightarrow 3$ transitions. Another suitable material might be p-InP having an energy gap of 1.35 eV and a split-off band at 0.11 eV below the valence band edge.

[21] Calculations on the field induced dichroism for the fundamental electronic absorption based on the drifted MB approach have been performed by Baumgardner, C. A., Woodruff, T. O.: Phys. Rev. **173**, 746 (1968).

However hot electron properties in p-type materials have widely been studied only in p-Ge and p-Si [8]. There is no such interest in p-type III–V materials mainly due to the fact that the minimum carrier concentrations which were obtained are still of the order of $10^{16} \, \text{cm}^{-3}$ and therefore also the mobilities are only about $1000 \, \text{cm}^2/\text{Vs}$ except for p-GaAs.

3.2.3. Radiative Recombination from Electric Field Excited Hot Carriers

In this and the following two sections we describe optical methods which have mainly been used to determine hot electron distribution functions in n-GaAs. This material was of particular interest due to the possibility of obtaining a bulk negative differential conductivity (BNDC) due to carrier transfer from the lowest lying Γ valley to X valleys in high fields. Although the optical measurements can only be performed up to fields below the threshold for the onset of the Gunn oscillations it was hoped to obtain useful information on the shape of the distribution functions. Beyond the threshold, the electric field is no longer homogeneous along the sample since high voltage domains are built up.

In this section we describe a method which is based on the observation of emission spectra caused by recombination radiation from electron-hole pairs. In an n-type material [153, 154] electron-hole pairs are created by photoionization using a light source (laser) operating at a photon energy higher than the band gap. In order to probe the electron distribution the intensity of the recombination radiation spectrum dependent on the photon energy is recorded[22]. Since the lowest conduction band at the Γ point has a curvature stronger than that of the highest valence band, the intensity of the recombination radiation spectrum at a given photon energy $\hbar\omega$ is approximately proportional to the number of electrons at $\hbar\omega - \Delta\varepsilon_g$. Thus, for a hypothetical semiconductor with flat valence bands, the recombination radiation spectrum would yield directly the electron distribution function. In a real semiconductor however both light and heavy holes have to be included in the analysis and also assumptions on the hole distribution have to be made. In Fig. 37 it is shown that the luminescence at a given photon energy results from two different transitions caused by the different curvature of the two valence bands. If all bands are assumed to be parabolic the occupancy O_i of the hole states at the same energy ε but at two different wave vectors k_1 and k_2 is given by [153]

$$O_i(k_i) = a^* \exp\left[-\alpha_i \varepsilon/(\alpha_c + \alpha_i) k_B T\right]$$

---

[22] The emission spectra reflect the hot carrier distribution function since the lifetime of the photoionized carriers is much larger than the relevant relaxation times within the bands. The additional electrons quickly reach a quasiequilibrium determined by the distribution function in the conduction band.

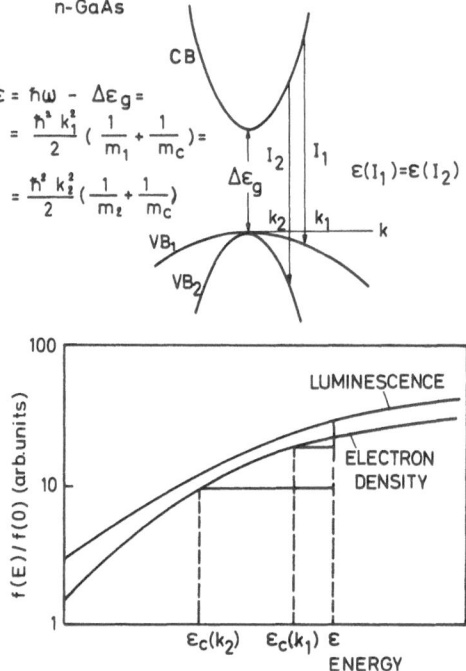

Fig. 37. Upper part: transitions of equal energy between light (VB_2, m_2), heavy hole band (VB_1, m_1) and the conduction band (CB, m_c). Lower part: ratio $f(E)/f(0)$ vs energy deduced from luminescence spectrum (Ref. [154])

where $\alpha_i = m_0/m_i$ ($i = 1, 2$ for Bands 1 and 2) and $\alpha_c = m_0/m_c$. a^* depends on the total hole density in thermal equilibrium. The total recombination radiation in a range $d\varepsilon$ will be proportional to the number of states multiplied by their occupancy in each band and also multiplied by the oscillator strength g_{ci}. The density of states factor will be proportional to $\varepsilon^{1/2}/(\alpha_c + \alpha_i)^{3/2}\, d\varepsilon$ (being proportional $k_i^2\, dk_i$). Thus the contribution to the luminescence $P_{l,i}$ for the transition I_1 or I_2 at a field E is given by

$$P_{l,i}(E, \varepsilon)\, d\varepsilon \sim f(E, \varepsilon_{ci}) \exp(-\varepsilon_{vi}/k_B T)\, g_{ci}\varepsilon^{1/2}/(\alpha_c + \alpha_i)^{3/2}\, d\varepsilon$$

with

$$\varepsilon_{ci} = \alpha_c \varepsilon/(\alpha_c + \alpha_i)$$
$$\varepsilon_{vi} = \alpha_i \varepsilon/(\alpha_c + \alpha_i)$$

where it is assumed that both hole populations are not heated by the field, and ε_{ci}, ε_{vi} are the energies from the band edges [153].

By summing over both contributions (from I_1 and I_2) and by taking the ratio to the zero field spectrum

$$\frac{P_l(E, \varepsilon)}{P_l(0, \varepsilon)} = \frac{\sum\limits_{i=1}^{2} g_{ci} f(E, \varepsilon_{ci}) \exp(-\varepsilon_{vi}/k_B T)/(\alpha_c + \alpha_i)^{3/2}}{\sum\limits_{i=1}^{2} g_{ci} f(0, \varepsilon_{ci}) \exp(-\varepsilon_{vi}/k_B T)/(\alpha_c + \alpha_i)^{3/2}}$$

with $f(0, \varepsilon_{ci}) \sim \exp(-(\varepsilon_{ci}/k_B T))$

$$\frac{P_l(E, \varepsilon)}{P_l(0, \varepsilon)} = \frac{\sum\limits_{i=1}^{2} g_{ci} f(E, \varepsilon_{ci})/(f(0, \varepsilon_{ci})(\alpha_c + \alpha_i)^{3/2})}{\sum\limits_{i=1}^{2} g_{ci}/(\alpha_c + \alpha_i)^{3/2}}. \tag{57}$$

According to Elliot's theory of absorption [152] electron-hole Coulomb interaction should be included. Therefore in Eq. (57) the effective mass factor should be replaced by $(\alpha_c + \alpha_i)^2$ both in the denominator and nominator. If the particular values of the effective masses are inserted, following relation results

$$P_l(E, \varepsilon)/P_l(0, \varepsilon)$$
$$= M(\alpha_c, \alpha_1) f(E, \varepsilon_{c1})/f(0, \varepsilon_{c1}) + M(\alpha_c, \alpha_2) f(E, \varepsilon_{c2})/f(0, \varepsilon_{c2}). \tag{58}$$

Usually the heavy hole mass contribution $M(\alpha_c, \alpha_1)$ is much larger than the light hole mass contribution $M(\alpha_c, \alpha_2)$. The ratio g_{c1}/g_{c2} is unity in materials having a valence band structure like GaAs [140].

In order to derive $f(E)/f(0)$ from the measured $P(E)/P(0)$ a graphical iterative procedure was made starting at $\varepsilon = 0$ [153]. The derivation presented here differs from that given in Ref. [154] where the different density of states factor for light and heavy hole was not included [153]. By expressing the experimental results in the form $P_l(E)/P_l(0)$ no corrections due to self absorption have to be made.

Experiments were performed by Southgate et al. [154–156] on n-GaAs at lattice temperatures of 77 K and 200 K and fields between 0–2000 V/cm. The GaAs samples were thin epitaxial layers having carrier concentrations of $1-5 \times 10^{15}$ cm^{-3}. A He–Ne laser was used to excite photoluminescence (Fig. 38). The spectrum was observed at the end of a 25 µs voltage pulse. The laser beam was chopped (200 Hz) and the voltage pulse was applied in the on and off periods of the beam in order to use a lock-in technique with the output signal from the photomultiplier.

The recombination radiation spectra in GaAs show with and without applied field an exciton peak approximately 4.5 meV below the band edge. The spectra exhibit a high energy tail observable between 1.52 and

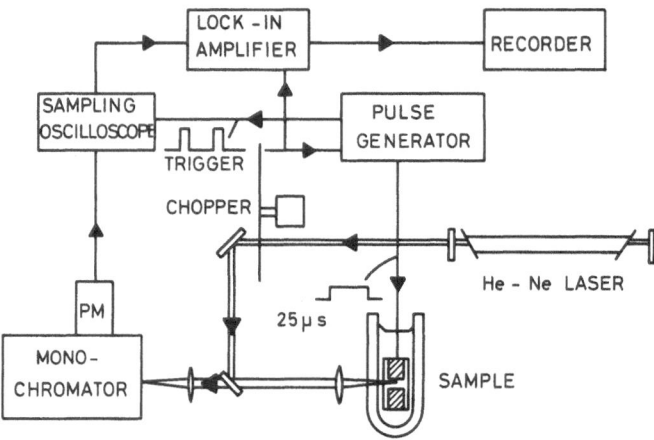

Fig. 38. Experimental arrangement for photoluminescence measurements. (After Ref. [154])

1.58 eV. In high fields the intensity of the photoluminescence increases at high energies due to the increasing number of high energy electrons. Near the band edge and below, the luminescence spectrum decreases due to the decreasing number of carriers at low energies. The decrease of the exciton peak can also be attributed to a decrease of exciton concentration. The Franz Keldysh effect [100], will cause only minor changes in the observed spectrum which are below the experimental resolution.

Southgate et al. [154] have deduced energy distribution functions for $T = 77$ and $T = 200$ K from the photoluminescence measurements. In Fig. 39 the experimentally derived distribution function normalized to the MB distribution of 77 K are compared with data based on calculations by Rees [121] and on a drifted MB distribution. A distinction between the applicability of the two models cannot be made, since the difference between the experimental data for samples having a concentration of 5.9×10^{14} and 4.3×10^{14} cm^{-3} are too large. It is suspected that the differences in the spectra of the samples are caused by inhomogeneity effects. In the calculation by Rees a kink at 34.5 meV, the optical phonon energy is present whereas the experiments yield a kink at 25 meV. At $T = 200$ K only a comparison with the drifted MB was possible showing at energies between 20 and 70 meV above the conduction band edge a small increase of the experimental $f(E)/f(0)$ above the values calculated for the drifted MB.

Measurements of the luminescence spectrum due to electron-hole recombination at acceptors have been performed by Salomon [157] in n- and p-InSb at 4.2 K. The data were however not published.

The experiments presented in this section have shown that photoluminescence measurements may be used to derive hot electron distribu-

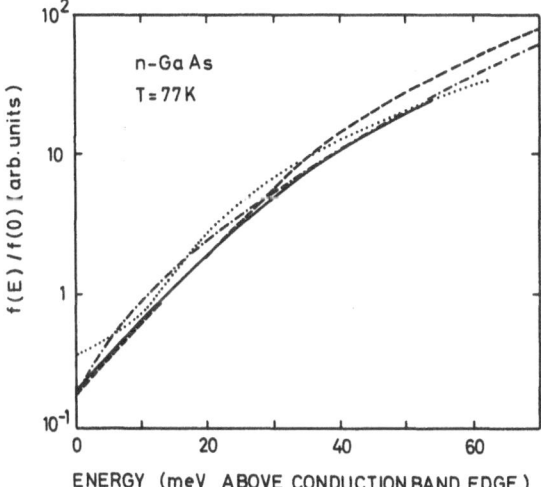

Fig. 39. Distribution functions for two GaAs samples (Ref. [154]). (······) $N = 4.3 \times 10^{15}$ cm^{-3}, (———) $n = 5.9 \times 10^{15}$ cm^{-3}, (– – –) calculation by Rees [121], (—·—·—) displaced MB distribution $v_d = 2.4 \times 10^7$ cm s^{-1}

tion functions up to 70 meV above the band gap. However until now the experimental resolution is insufficient to distinguish between different theoretical models. It is remarkable that the experimentally deduced distribution functions show no kink at the optical phonon energy in GaAs. In addition, in a preliminary attempt to deduce hot-carrier distribution functions in Ge at $T = 4.2$ K, indirect band-gap recombination radiation was analysed in fields up to 35 V/cm, and interpreted as evidence for MB distributions with T_e up to 26 K [158]. The anisotropy of hot-carrier recombination radiation in n-InSb at $T = 85$ K and fields up to 400 V/cm was investigated in Ref. [159], unfortunately under spatially nonuniform conditions.

3.2.4. Optical Interband Absorption in Degenerate Materials

In Section 2.2.8 a method for determining electron temperatures was described, based on the shift of the electronic fundamental absorption edge to higher photon energies in a degenerate semiconductor. A refinement of this method with respect to spectral resolution, is capable of determining electron distribution functions.

The fundamental absorption in a degenerate n-type material at a certain photon energy is proportional to the number of occupied states in the conduction band at an energy which is determined by the photon energy and by the band structure. The number of unoccupied states is given by $(1 - f(\varepsilon))$, where $f(\varepsilon)$ denotes the distribution function in the conduction band (see Fig. 15). Neglecting warping of the valence bands, an unique relation between the electron energy in the conduction band and the absorbed photon energy is established for direct transitions. Since K, the absorption coefficient, is directly proportional to $(1 - f(\varepsilon))$, the measured change of the optical interband absorption allows the evaluation of the distribution function in high electric fields.

In deducing a distribution function from the absorption edge measurements the following effects have to be considered which might also influence the field induced change of the absorption coefficient [136]:

a) Beside the direct transitions an exciton line may occur.

b) At low values of the absorption constant (up to $500\,\mathrm{cm}^{-1}$ in n-GaAs) K is dependent on the impurity content. Often an exponential tailing of the absorption for $\hbar\omega < \Delta\varepsilon_g$ is found (Urbach region) [160].

c) The Franz Keldysh effect will modulate the absorption [161].

d) Indirect absorption processes involving absorption or emission of phonons [162].

In the experimental investigations by Jantsch and Heinrich [163], n-GaAs having a carrier concentration of $1 \times 10^{18}\,\mathrm{cm}^{-3}$ was investigated for fields up to $450\,\mathrm{V/cm}$. At this high carrier concentration the exciton absorption disappeares due to screening of the Coulomb interaction. Further the Franz Keldysh effect is negligible in the range of fields investigated in [163]. Experimentally due to the Franz Keldysh effect in semi-insulating GaAs a change of K of about $2\,\mathrm{cm}^{-1}$ at $2\,\mathrm{kV/cm}$ is observed [164]. McGroddy and Christensen [165] have made an estimate on the contribution of indirect absorption to the observed K in GaAs. A change of indirect absorption may occur due to the heating of the carriers, which will be able to emit optical phonons for energies in excess of $\hbar\omega_0 = 34.5\,\mathrm{meV}$. From the estimate [165] a change of K of about $20\,\mathrm{cm}^{-1}$ may result at $1\,\mathrm{kV/cm}$. Thus below $500\,\mathrm{V/cm}$, K due to this effect will be smaller than $5\,\mathrm{cm}^{-1}$.

Monochromatic, linearily polarized light was used in the experiments [163], the direction of polarization being parallel or perpendicular to the applied electric field. The electric field was applied in pulses of 200 ns duration perpendicular to the direction of light which was detected by an avalanche photodiode. The rise time of the detector limited the range of electric fields in order to ensure a constant lattice temperature.

The distribution function was obtained from a comparison of $K_E(\hbar\omega)$ with $K_{\mathrm{th}}(\hbar\omega)$ according to

$$(1 - f(E, \varepsilon)) = (K_E(\hbar\omega)/K_{\mathrm{th}}(\hbar\omega))(1 - f(0, \varepsilon)). \tag{59}$$

A complication in the analysis presents the dependence of the energy gap on carrier concentration. As shown in Fig. 40 the absorption con-

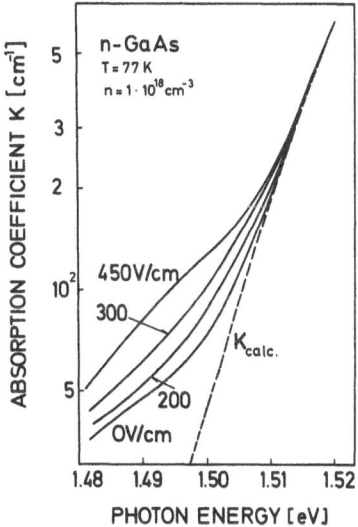

Fig. 40. Influence of electric fields up to 450 V/cm on the electronic fundamental absorption in *n*-GaAs [163]

stant under ohmic conditions deviates considerably from a calculated absorption coefficient using the formula for direct allowed transitions and Fermi statistics for parabolic bands. A value of 1.47 eV was used for $\Delta\varepsilon_g$ in agreement with calculations by Hwang [166] who obtained also a shift of the gap to smaller energies at high carrier concentrations.

The application of fields increases the absorption constant in the range of photon energies from 1.48–1.51 eV.

At higher photon energies $(1 - f(\varepsilon))$ is already appreciable different from 1 and many empty states exist in the conduction band $(\varepsilon \approx \varepsilon_F)$ so that the changes between the high field-$f(E)$ and thermal equilibrium distribution function $f(0)$ are too small to be detectable in the absorption coefficient due to allowed transitions.

The deviations of K under ohmic conditions in the low energy region (< 1.51 eV) from the calculated values arise due to the altered density of states at the bottom of the conduction band [166]. However in this region, where the influence of the band tails is evident, the distribution function has to be deduced. If only transitions from the heavy hole band are considered, then at a photon energy of 1.51 meV the initial state in the heavy hole band is only 5 meV below the valence band edge. Therefore this energy will be even smaller at smaller photon energies and so the deviations of the conduction band from parabolic band shape are not serious in the interpretation of the results.

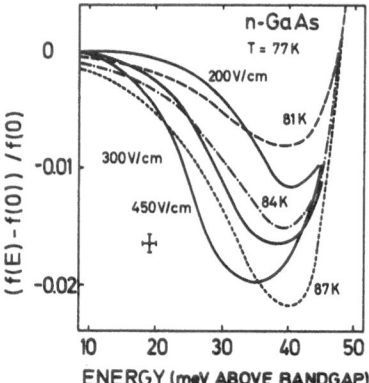

Fig. 41. Change of distribution function in electric fields from 200–450 V/cm in n-GaAs at 77 K, $n = 1 \times 10^{18}$ cm^{-3}. (After Ref. [163]). ✛ indicates experimental resolution

Using Eq. (59) and inserting the Fermi distribution for $f(0)$, the resulting relative change of distribution functions due to the application of electric fields was calculated. The results are represented in Fig. 41. The minimum in $\Delta f(E)/f(0)$ shifts to lower energies when the electric field in increased. In Fig. 41 also calculated distribution functions are included using Fermi distribution functions at 81, 84, and 87 K. The comparison shows that the population at low energies (20–25 meV) is higher than expected by the Fermi distributions whereas it is lower for higher energies above the gap. The method of the electro-absorption in the fundamental absorption edge has until now only be applied to n-GaAs but there are no obstacles to use it in other semiconductors. It is capable of detecting small changes (Fig. 41) of the distribution function in high fields but is restricted to degenerate semiconductors. No numerical solutions of the BE applicable to degenerate statistics are until now available although it was pointed out that in principle the iterative methods should be appropriate for this problem [11, 120].

3.2.5. Inelastic Light Scattering from Hot Electrons

In a phenomenological description, inelastic light scattering is caused by density fluctuations, if these are connected with changes of the polarizability. The scattering cross section yields information on the spectrum of the density fluctuations, which is manifested in a dependence on frequency and wave vector. If the free carriers contribute to the polarizability, a change of free carrier properties due to some external perturbation, will influence the scattered spectrum.

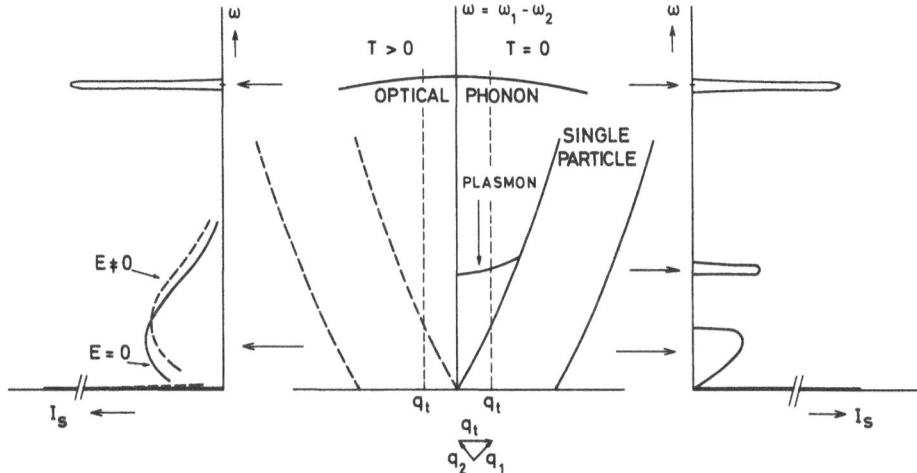

Fig. 42. Sketch of elementary excitations in solids. For $T = 0$ single particle spectrum is confined by the magnitude of q_t and k_F. At higher temperatures (lefthand side) the Fermi surface becomes smeared and more electrons participate in the scattering process. q_t is the momentum transfer induced by the exciting laser (Ref. [172]). I_S denotes the scattered intensity and E the electric field

Charge density fluctuations may be caused by the individual moving electrons in phase space (single particle excitations) and by collective phenomena, e.g. plasmons. The part of the scattered spectrum dominated by the single particle excitations will yield information on the velocity distribution of the electrons [167].

In Fig. 42 the excitations contributing to the scattered intensity are sketched. Here we are mainly interested in the electronic contribution. q_t denotes the momentum transfer imparted by the laser. For charge density fluctuation scattering the single particle excitation spectrum will dominate over the collective part if $q_t > q_D$, the screening wave vector of the electron gas. The momentum change q_t is determined by the frequency of the incident light, of the scattered light and by geometry. In the case of 90° geometry $q_t \approx \sqrt{2} q_1$, q_1 being the wave vector of the incident light. In the scattering process the energy $\hbar \omega$ which yields the frequency shift is determined by [172]

$$\hbar \omega = (\hbar k + \hbar q_t)^2 / 2m - \hbar^2 k^2 / 2m .$$

For $T = 0$ the possible ω, q_t combinations for scattering events are shown in the righthand side of Fig. 42. The shape of the spectrum at $T = 0$ depends beside the density of states on the magnitude of q_t compared to the Fermivector k_F: only if $q_t > 2k_F$ all electrons of the Fermisphere can take part in the scattering process. At high temperatures, in a MB approximation this condition will be not so restrictive. On the lefthand side of Fig. 42 the scattered intensity is sketched for $T \neq 0$ and the influence of an electric field is indicated. By choosing a particular laser line and the geometry, q_t is fixed. Therefore the frequency shift ω of the scattered radiation yields information on the electron energy. The scattering cross section depends on the occupation of the initial and final states. The frequency dependence of the scattering intensity thus reflects the distribution function.

The calculation of the scattering cross section due to single particle excitations was based originally on the model of charge density fluctuations [167–169]. However due to the fact that the electrons in a semiconductor have an energy-momentum relation which differs from that of the free electron other mechanisms are also important. Hamilton and McWorther [170] have calculated a scattering cross section due to electron spin-density fluctuations. These are coupled to the electromagnetic radiation via the spin orbit effects, taking into account the real structure of the valence bands. The calculation of the Raman cross section involves first and second order perturbation theory. We only quote the result obtained by Mooradian and McWorther [171] for low carrier concentrations where $q_D \ll q_t$ and a distribution function $f(k) \ll 1$. The scattering cross section $d^2\sigma/d\Omega\,d\omega$ is given by

$$d^2\sigma/d\Omega\,d\omega = (\omega_2/\omega_1)\,(m/q_t)\,(e^2/m_0c^2)^2\,(2/8\pi^3)\int dk_x\,dk_y$$
$$\cdot G^2(\hbar k_x, \hbar k_y, m\omega/q_t)\,f(\hbar k_x, \hbar k_y, m\omega/q_t) \tag{60}$$

where

$$G(k) = 1 + \frac{2P_m^2}{3m_0}\sum_{i=1}^{3}\frac{\varepsilon_{gi}}{\varepsilon_{gi}^2 - ((\hbar\omega_1 + \hbar\omega_2)/2)^2}.$$

ω_1 and ω_2 are the incident and scattered frequencies and P_m is the interband momentum matrix element, ε_{gi} are the energy differences between the conduction band and the valence bands ($i = 1 - 3$) at k. The coordinate system was chosen such that $q_t \parallel z$. If G^2 can be taken outside of the integral, the scattering cross-section at a frequency shift ω will be proportional to the number of carriers with velocity $v_z = \omega/q_t$. Thus the velocity distribution of the carriers may be determined.

Mooradian [172] has shown that the observed scattering cross-section of n- and p-GaAs at $T = 300$ K by exciting single particle excitations with a YAG–Nd^{3+} laser is in good agreement with a calculation using a MB distribution. Therefore, in later investigations [171, 173] it was tried to deduce distribution functions in high electric fields in n-GaAs.

The experimental arrangement is shown in Fig. 43. A Q-switched YAG–Nd^{3+} laser operating at 1.06 μm was used as a light source. A

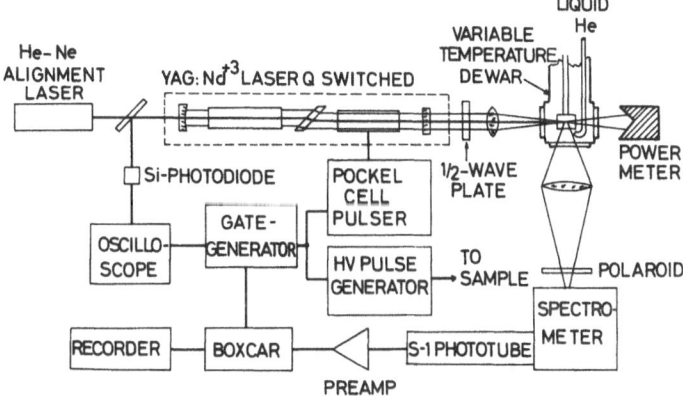

Fig. 43. Experimental arrangement for inelastic light scattering of hot electrons. (After Ref. [171])

GaAs sample, having a carrier concentration of about $4 \times 10^{15} \, \text{cm}^{-3}$ was immersed in liquid He at 1.6 K. The scattered light, detected at 90° from the direction of the incident beam, was collected from the central part of the sample and focused to the slit of the spectrometer. The electric field was applied in pulses of 50 ns duration and a repetition rate of 50 Hz. During the light pulse the sample was heated up to 25 K as determined from the scattered spectrum without applying the electric field. The laser beam also excited additional electrons into the conduction band, altering the free carrier density by about 10 %.

Measurements of the field induced change of the scattering cross-section were only performed at low temperatures. At 300 K, no change of $d^2\sigma/d\Omega \, d\omega$ was observable up to 2000 V/cm since the thermal velocity was already quite high ($v_{\text{th}} = 4 \times 10^7$ cm/s). At larger frequency shifts also phonon scattering overlapped with the spectra from the excitation of the electrons.

At low temperature the measurements showed changes of the scattering cross-section with electric field. The anti-Stokes components of the spectrum were analyzed. For electron temperatures in excess of 50 K the plasmon contribution ($\tilde{v}_p = 20 \, \text{cm}^{-1}$) to the spectrum is severely damped since then $q_t > q_D$. Figure 44 shows results for $E = 1400$ V/cm.

Since the scattering cross-section is a functional of the distribution function, no direct data on this function can be extracted from this experiment. Mooradian and McWorther [171] did not use a distribution function with several parameters to fit the experimental data. They calculated anti-Stokes scattering cross-sections for a drifted MB distribu-

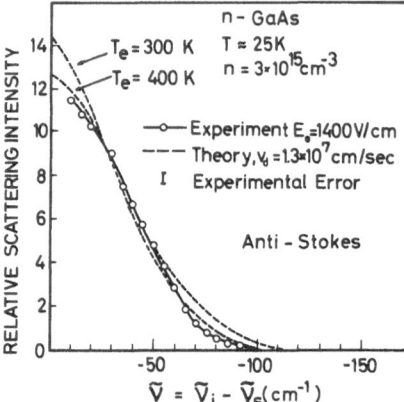

Fig. 44. Inelastic light scattering spectrum of n-GaAs in a field of 1400 V/cm. Experimental data are compared with calculations for two displaced MB distributions at 300 and 400 K (Ref. [171])

tion. In the experimental set-up the static field E was applied parallel to the propagation direction of the incident light and the scattered light was observed under 90°. For charge density fluctuation scattering dominating [171], $d^2\sigma/d\Omega\,d\omega$ is given by

$$d^2\sigma/d\Omega\,d\omega \approx (\omega_2/\omega_1)\,(n/q_t)\,(e^2/m_0 c^2)^2\,(m/2\pi k_B T_e)^{1/2}\,G^2\,(v_d/\sqrt{2},0,\omega/q_t)$$

$$\cdot \exp\left[-\frac{m}{2k_B T_e}\,(\omega/q_t + v_d/\sqrt{2})^2\right] \tag{61}$$

where for G^2 a value at the extremum of the integrand was used and it was taken outside of the integral.

The experimental limitations are caused by the limited resolution near the laser line (3 cm^{-1} in the experiment without electric fields) and the smearing and small single particle intensity at large shifts $\omega_1 - \omega_2$ and high electric fields. However the authors state that the determined scattering cross-section definitely deviates from that calculated for a MB distribution and that this deviation is outside the experimental error. No attempt was made to use a distribution function determined by Monte Carlo or iterative calculations for an evaluation of the scattering cross-section.

At the end of this section, in Table 3 the experimental data are summarized which have yielded information on hot electron distribution functions.

4. Comparison of Experimental and Numerical Results — Conclusions

4.1. Electron Temperature Model

We will first discuss calculations applicable to the methods described in Sections 2.1.1–2.1.5. In these sections experiments were described which were performed in a range of lattice temperatures and electric fields in which ionized impurity scattering dominates the momentum relaxation. In this case a momentum relaxation time exists. The mobility $\mu = (e/m)\langle\tau\rangle$ is then a function of T_e only and Eq. (30d) is inserted for the momentum relaxation time [174]. The dependence of the electron temperature T_e on the electric field E is calculated from the energy balance equation. In the energy balance equation $e\mu E^2 = \langle d\varepsilon/dt\rangle = P(T_e, T)$ the energy loss rate, has to be calculated which is now described.

The rate of change of carrier energy because of phonon scattering is obtained by substracting the energy gain by absorption of phonons from the energy loss by emission of phonons. The gain is given by $\hbar\omega_q$ times the probability $W(k, k + q)$ of a phonon being absorbed, and the loss is given by $\hbar\omega_q W(k, k - q)$, $W(k, k - q)$ is the probability for phonon emission. The expressions have to be summed over all possible q yielding

$$d\varepsilon/dt$$
$$= (- V/8\pi^3)\left\{\int_{q_1^-}^{q_1^-} \hbar\omega_q W(k, k - q)\, d^3 q - \int_{q_1^+}^{q_1^+} \hbar\omega_q W(k, k + q)\, d^3 q\right\}. \tag{62}$$

In order to obtain the average energy loss rate per carrier this expression has to be averaged over the distribution function

$$P = \langle d\varepsilon/dt\rangle = (2/8\pi^3 n)\int f(k)\, (d\varepsilon/dt)\, d^3 k. \tag{63}$$

Generally the integration can only be performed numerically since $d\varepsilon/dt$ is a complicated function of k. For some special cases P is listed in Table 4. Since the parameter of f is the electron temperature, the energy balance equation relates the applied field E to $T_e (e\mu E^2 = P)$.

For degenerate statistics in Eq. (62) a factor $(1 - f(\varepsilon \mp \hbar\omega_q))$ has to be included to ensure that the state in which the electron is scattered is not occupied. The limits of the q integration are derived from the δ-functions which occur in the transition probabilities $(\delta(\varepsilon(|k \mp q|) - \varepsilon(|k|) \pm \hbar\omega_q))$ and conserve energy. In the range of carrier concentrations investigated, often screening by free electrons has to be taken into account. The effect of static screening on the electron phonon interaction reduces the interaction potentials by a factor $\lambda_D^2 q^2/(\lambda_D^2 q^2 + 1)$ where λ_D

Table 3. Determination of hot carrier distribution functions

Method	Material	Carrier conc. (cm^{-3})	T (K)	E (V/cm)	Results	Ref.
Intervalence band absorption	p-Ge	4.4×10^{15}	93	0–250	$f(\varepsilon_1)$: MB distribution, T_e: 93–115 K, no T_e definable	[126]
			293	1480		
	p-Ge	6×10^{14}	93	0–350	$f(\varepsilon_1)$: approx. MB distribution T_e: 93–139 K	[127]
		3.9×10^{15}	93	0–350	$f(\varepsilon_1)$: approx. MB no T_e given	[127]
		9.1×10^{15}	93	0–350	$f(\varepsilon_1)$: approx. MB T_e: 93–108 K	[127]
	p-Ge	1.9–10^{15}	77	130–2150 ($E \parallel [111]$, $E \parallel [100]$)	$f(\varepsilon_1)$ $\varepsilon_1 = 0.015$–0.06 eV	[129]
Dichroism of intervalence band absorption	p-Ge	5.9×10^{14}	77	760 ($E \parallel [100]$)	$f_0(\varepsilon_1), f_2(\varepsilon_1)$ $\varepsilon_1 = 0.015$–0.055 eV	[130]
	p-Ge	6×10^{14}	77	350 ($E \parallel [100]$)	$f_0(\varepsilon_1), f_2(\varepsilon_1)$ $\varepsilon_1 = 0.010$–0.060 eV $T_1 = 91$ K, $T_2 = 77$ K	[131]
	p-Ge	2.77×10^{15}	77	436 695 926	$f_2(\varepsilon_1)/f_0(\varepsilon_1)$, $\varepsilon_1 = 0.01$–0.05 eV	[132]
	p-Ge	$\approx 8 \times 10^{15}$	77	500	$f_0(\varepsilon_1), f_2(\varepsilon_1), f_0(\varepsilon_2)$ $\varepsilon_1, \varepsilon_2 = 0.01$–0.055 eV	[147]

Effect	Material	Concentration (cm^{-3})	T (K)	Field / Stress	Distribution function	Ref.
	p-Ge	6.5×10^{14}	85	380	$f_0(\varepsilon_1)$, $\varepsilon_1 = 0.01$–0.055 eV $T_1 = 170$ K, $T_2 = 91$ K	[133]
				960	$f_0(\varepsilon_1)$ $T_1 = 200$ K, $T_2 = 106$ K	
				1710	$f_0(\varepsilon_1), f_2(\varepsilon_1)/f_0(\varepsilon_1)$ $T_1 = 214$ K, $T_2 = 140$ K	
	p-Ge	2.1×10^{15}	85	250 500 1000 ($E \parallel [111]$)	parametrized drifted MB $f(\varepsilon_1) = f(T_\parallel, T_1, T_2, v_d)$	[136]
Dichroism of intervalence band absorption under uniaxial stress	p-Ge $X = 5900, 8850,$ $11\,800$ kp/cm^2	2.1×10^{15}	85	250 500 1000 ($E \parallel [111]$)	Parametrized drifted MB $f(T_\parallel, T_1, T_2, v_d)$	[136]
Luminescence (light-heavy hole transition)	p-Ge	6×10^{14}	85	4000 5200 5800	$f(\varepsilon_2), f_2(\varepsilon_2)/f_0(\varepsilon_2)$	[150] [151]
Radiative (electron-hole) recombination	n-GaAs	$\left.\begin{array}{l} 4.3 \times 10^{15} \\ 5.9 \times 10^{15} \\ 4.3 \times 10^{15} \end{array}\right\}$	77 200	≈ 1500 ≈ 2500	$f(\varepsilon) = 0$–0.07 eV $f(\varepsilon) = 0$–0.07 eV	[154]
Electronic fundamental absorption edge	n-GaAs	1×10^{18}	77	200 300 450	$f(\varepsilon) = 0$–0.05 eV	[163]
Inelastic light scattering	n-GaAs	3×10^{15}	≈ 25	1400	Anti Stokes spectrum no MB distribution	[171]

$f(\varepsilon_1)$ heavy hole distribution function.
$f(\varepsilon_2)$ light hole distribution function.
$f_0(\varepsilon_1), f_0(\varepsilon_2)$ first term in expansion in Legendre polynomials.
$f_2(\varepsilon_1), f_2(\varepsilon_2)$ third term in expansion in Legendre polynomials.
T_1, T_2 "carrier temperature" for $\varepsilon \lessgtr$ optical phonon energy.
$f(\varepsilon)$ electron distribution function.

Table 4. Energy loss rates for parabolic bands

A) Non degenerate statistics	P	Assumptions	Ref.
Acoustic phonon (Deformation potential)			
Simple model	$8\sqrt{2}\,\pi^{-3/2}(E_1^2\, m^{5/2}/\hbar^4 d)(k_B T_e)^{3/2}(1 - T/T_e)$	$N_q = k_B T/\hbar\omega_q - \frac{1}{2}$	[10]
Many valley model	$8\sqrt{2}\,\pi^{-3/2}(\Xi_0^2\, m_t^2\, m_l^{1/2}/\hbar^4 d)(k_B T_e)^{3/2}(1 - T/T_e)$ $\Xi_0^2 = \Xi_d^2[\frac{2}{3} + \frac{1}{3}(m_{tl}/m_t)(\Xi_u/\Xi_d + 1)^2]$		
Optical phonon			
Simple model	$\sqrt{2/\pi}(D^2 m^{3/2}/\pi\hbar^2 d)(k_B T_e)^{1/2}\,\dfrac{e^{x_0 - x_e} - 1}{e^{x_0} - 1}\,(x_e/2)\,e^{x_e/2}\,K_1(x_e/2)$ $x_e = \hbar\omega_0/k_B T_e \qquad x_0 = \hbar\omega_0/k_B T$		[10]
Many valley	$\sqrt{2/\pi}(D^2 m_t m_l^{1/2}/\pi\hbar^2 d)(k_B T_e)^{1/2}\,\dfrac{e^{x_0 - x_e} - 1}{e^{x_0} - 1}\,(x_e/2)\,e^{x_e/2}\,K_1(x_e/2)$		[10]
Intervalley	as optical phonon: $\hbar\omega_0 \to \hbar\omega_{i,j} \qquad D \to D_{i,j}$		

B) Degenerate statistics	P	Assumptions	Ref.
Acoustic phonon			
Deformation potential	$(2m)^{5/2}\, k_B^{3/2}\, E_1^2\,(\pi d\hbar^4)^{-1}\, T_e^{1/2}(T_e - T)\, F_1(\eta)/F_{1/2}(\eta)$	$(8mu^2\eta/k_B T)^{1/2} \ll 1$	[177] [178]
Piezoelectric coupling	$\frac{16}{5}\pi(e\,e_{14})^2(2m)^{3/2}\, k_B^{1/2}\, h^{-2}\, x^{-2}\, d^{-1}(T_e - T)\, T_e^{-1/2}\, F_0(\eta)/F_{1/2}(\eta)$		[178]
With Thomas-Fermi screening	as above but: $F_0(\eta) \to G_0(\eta) = \int_0^\infty x^2(x + x_D)^{-2}(1 + e^{x - \eta})^{-1}\,dx$ $x_D = \frac{1}{16}\pi n e^2 h^2 m^{-1} x^{-1}(k_B T_e)^{-2}\, F_{-1/2}(\eta)/F_{1/2}(\eta)$		[27] [36] [45]

Polar optical phonon

$$\frac{\sqrt{2}eh\omega_0}{(mk_BT_e)^{1/2}F_{1/2}(\eta)}(meh\omega_0/\hbar^2)(x_\infty^{-1}-x^{-1})\int_y^\infty f(x)[1-f(x-y)]$$
$$\cdot \ln[(x/y)^{1/2}+(x/y-1)^{1/2}]\,dx \qquad\qquad T \ll \theta_0 \qquad [45]$$

$$y=\theta_\omega/T_e \qquad \hbar\omega_0=k_B\theta_0 \qquad f(x)=1/(\exp(x-\eta)+1)$$

Simple model: isotropic effective mass

E_1 Acoustic deformation potential constant.
d Mass density.
Ξ_d, Ξ_u Deformation potential constants for ellipsoidal energy surfaces.
m_l, m_t Longitudinal and transverse mass in ellipsoidal model.
D Coupling constant for nonpolar optical modes.
K_1 Bessel function of second kind with imaginary argument.
$\hbar\omega_0$ Optical phonon energy.
D_{ij} Intervalley coupling constant for scattering from the i-th valley to the j-th.
$\hbar\omega_{ij}$ Intervalley phonon energy.
$F_k(\eta)$ Fermi integrals.
e_{14} Piezoelectric coupling constant.
x Static dielectric coupling constant.
x_∞ Optical dielectric coupling constant.
θ_0 Optical phonon Debye temperature.

is the screening length usually calculated using the Thomas-Fermi expression [175, 176][23].

If several scattering mechanisms occur, the energy loss rate is the sum of all contributions. In calculating the mobility, the sum of the reciprocals of the individual relaxation times which contribute to momentum relaxation yields the total $1/\tau$ which is then averaged over the distribution function.

In the Sections 2.1.1–2.1.5 we were mainly concerned with III–V compounds at low temperatures. For the energy loss mechanisms acoustic phonon scattering both via the deformation potential and the piezoelectric coupling and polar optical phonon scattering have to be considered. For the momentum relaxation, ionized impurity scattering dominates all other scattering mechanisms at low fields. The upper limit of the electric field depends on the particular material investigated and on the carrier concentration and ranges from $300\,\text{mV/cm}$ for n-InSb ($n \approx 10^{14}\,\text{cm}^{-3}$) to $30\,\text{V/cm}$ for n-GaSb ($n \approx 10^{18}\,\text{cm}^{-3}$). At higher fields polar optical scattering, for which no momentum relaxation time can be defined, influences the momentum relaxation. Then the method of Fröhlich and Paranjape [15] should be used. Since small gap materials like InSb exhibit a strong non-parabolicity this should be taken into account[24]. The band shape not only influences the density of states but also alters the scattering probability [179] since a p-like admixture to the electron wave functions has to be included. The expressions for the energy loss rate can only be evaluated numerically if non-parabolicity and degeneracy are considered. Integral expressions have been compiled for the above mentioned scattering processes in Ref. [76].

A main objective of the low temperature hot electron investigations in III–V compounds was the determination of the deformation potential constants. This quantity was used as a fitting parameter in order to relate the experimentally observed energy loss rate $e\mu E^2$ to a calculated P at a certain electron temperature. However deformation potential constants ranging from $5\,\text{eV}$ to $30\,\text{eV}$ [29, 37] have been reported for n-InSb as deduced from hot electron experiments and even a dependence of this constant on carrier concentration was anticipated [37]. It may be suspected that the formalism of the electron phonon interaction used in the evaluation of transition probabilities based on adiabatic principle and time dependent perturbation theory is not applicable at low temperatures. The adiabatic principle [180] fails to be acceptable when the period of a lattice wave is shorter than the time τ between successive

---

[23] The effect of screening on the phonon dispersion can usually be neglected in III–V compounds.

[24] See e.g. Costato, M., Reggiani, L.: J. Phys. C (Solid State Phys.) **5**, 159 (1972) for calculations on deformation potential and polar scattering.

collisions of an electron which means that the condition $\hbar/(k_B T) < \tau$ is not satisfied. According to Peierls [181] the use of the BE and of perturbation theory are not justified unless this inequality is satisfied. However for degenerate statistics $k_B T$ can be replaced by ε_F which makes this condition less restrictive [12, 182]. Related to this condition [180] is also $1 < k l_e$, which states that the mean free path of an electron l_e between successive collisions should be greater than its wavelength: the wave function of an electron should last at least a whole oscillation before it can be assigned a whole wavelength. Although these difficulties were already known for a long time, only Maneval et al. [26] and Tsidilkovskii and Demchuk [183] have considered them. Using degenerate samples these requirements are easier to fulfill. The deformation potential constants deduced from hot electron measurements in degenerate samples (Bauer and Kahlert [76]) of n-InSb with carrier-concentrations ranging from 5.9×10^{15}–6.9×10^{16} cm^{-3} did not vary with concentration. A value of 6.9 ± 0.4 eV was obtained in good agreement with Ehrenreich's value of 7.2 eV [179] obtained from the pressure dependence of the energy gap (for a discussion see [76]).

The dependence of the energy loss rate P on T_e, typical for several III–V compounds (InSb, InAs, GaSb) is shown in Fig. 45. At low temperatures accoustic phonon scattering determines P whereas at high temperatures the steep increasing polar optical phonon scattering takes over. Thus also the dependence of $T_e(E)$ is understandable (Fig. 2). Up to 10–12 K the electron temperature increases rapidly with field since acoustic phonon scattering is not very effective. At higher temperatures however polar optical phonon scattering dominates the energy loss process and the increase of T_e with E is severely diminished. There is however a temperature range in which the experimentally observed energy loss rate is higher than the calculated acoustic energy loss rate at temperatures where polar optical phonon scattering is not yet responsible for the energy dissipation. Dependent on the carrier concentration, it ranges from about 12 K–20 K (InSb, GaSb). This behaviour was observed several times [28, 31, 37, 75, 76]. Stradling and Wood [184] have suggested an additional scattering process, the emission of two transverse acoustic phonons $(2TA)$ of opposite wave vector having an energy corresponding to the zone edge of the Brillouin zone in the X point[25]. In hot electron magnetophonon experiments [184, 185] an oscillation series was found with just a characteristic energy of 10.3 meV which is exactly the energy of $2TA$ phonons at the X point in InSb. The temperature dependence of the energy loss rate between 12 and 20 K

[25] Recently Baumann, K.: Acta Phys. Austriaca **37**, 350 (1973) has made a comparison between the $2TA$ mechanism and the LO mechanism based on a pseudopotential calculation. His estimate yields results in qualitative agreement with the experimental data.

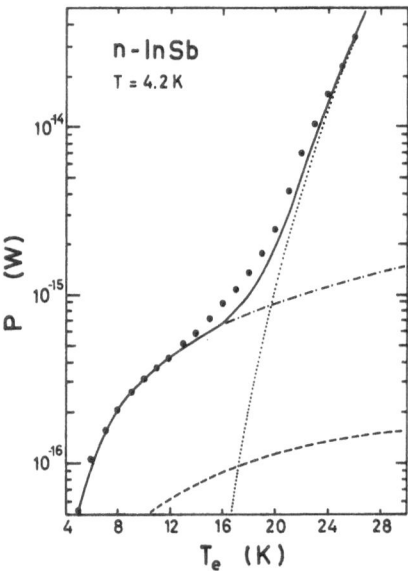

Fig. 45. Energy loss rate in n-InSb at 4.2 K ($n = 6.9 \times 10^{16}$ cm^{-3}). ● deduced from SdH experiments, (– – –) loss rate due to piezoelectric scattering ($e_{14}^2 = 5.62 \times 10^8$ dyn/cm^2), (— · — · —) loss rate due to sum of piezoelectric and deformation potential scattering ($E_1 = 6.9$ eV), (·····) polar optical phonon scattering ($\theta_0 = 278$ K), (——) sum of all contributions; calculations include non-parabolicity (Ref. [76])

can be explained, if such a process is considered to be similar to the polar optical phonon emission. The two phonon mechanism may have a much smaller transition probability and will be still effective [184]. Just by comparing the number of electrons which are able to emit $2TA$ phonons to those capable of emitting one optical phonon a factor $\exp((\hbar\omega_0 - 2\hbar\omega_{TA})/k_B T_e)$ results (≈ 2700 for $T = 20$ K). In addition the process is favoured by the high density of states of the TA branch near the zone edge. However until now there is no matrix element for this transition available. It may be worth to mention that a $2TA$ process in InSb with phonons from the X point has been observed in far infrared transmission experiments [186].

It was already mentioned that experiments on the Hall coefficient and on the magnetoresistence performed in low doped n-InSb are difficult to compare with theoretical predictions due to the change of carrier concentration in the magnetic field (magnetic freeze-out). Often the electron temperature concept is abandoned and it is tried to derive hot electron distribution functions if quantizing magnetic fields are considered [57, 58, 187]. However both the methods based on calculated distribution functions [187] as well as on the displaced MB approach

[188, 189] fail to yield a quantitative description of the hot electron magnetophonon effect. This effect results from a resonant interaction of electrons in Landau levels with polar optical phonons and leads to resistance oscillations with magnetic field. The calculations predict too large oscillatory amplitudes which may result from the improper handling of the Landau level broadening.

Calculations based on the electron temperature model yielded good agreement with the increase of T_e with E in InAs, InSb and GaSb [74–76] deduced from SdH measurements. Especially deformation potential constants E_1 (n-InSb: 6.9 ± 0.4 eV, n-InAs: 4.05 ± 0.25 V, n-GaSb: 5.0 ± 0.5 eV) were deduced which are in agreement with values found from an analysis of the temperature dependence of the mobility [190]. Calculations similar to that performed by Yamada [187] for non-degenerate statistics have until now not been performed for degenerate statistics. No exact theory of the energy loss rate of a degenerate electron gas has been developed so far for a quantizing magnetic field under conditions were several Landau levels are occupied. The experimental SdH data ($B \parallel j$), however, indicate that the "electron temperature" does not depend significantly on the magnetic field under these conditions. Recently Kocevar [78, 191] has derived the energy loss rate for degenerate statistics, taking into account several Landau levels and associated change of scattering rates within the frame work of the electron temperature model. His expressions for the energy loss rate explain qualitatively the small variation of T_e with B observed in SdH measurements in InSb [76]. A resonant enhancement of the energy loss occurs due to resonant emission of polar optical phonons at magnetic fields where the energy separation between adjacent Landau levels equals the polar optical phonon energy or the energy associated with the $2TA$ process. However the inclusion of the Landau level collision broadening which will be very important as far as the quantitative variation of T_e is considered, was until now only incorporated by the Dingle temperature.

In Section 2.1.5 we have presented the time dependent increase of T_e with E. For the calculation of the time dependent mobility change or for the time dependence of the SdH oscillations amplitudes it is useful to define a differential energy relaxation time

$$\tau'_\varepsilon = d\langle\varepsilon\rangle/dP = e^{-1} d\langle\varepsilon\rangle/d(\mu E^2).$$

For an energy independent energy relaxation time this definition is equivalent to Eq. (25). For degenerate semiconductors τ'_ε is defined according to

$$1/\tau'_\varepsilon = (1/c_h(T_e)) dP(T_e, T)/dT_e$$

where $c_h(T_e)$ is the electronic specific heat.

In Sections 2.1.6–2.1.9 we have described investigations performed with many valley semiconductors. The balance equations can be applied to these materials too and have been used to calculate beside electron temperatures the number of carriers in the different valleys. Since the detailed calculations are described elsewhere [8, 192] we only outline the principles. Usually it is assumed that in each valley a MB distribution with generally a different electron temperature exists. In order to calculate this temperature the energy balance equation for the i-th valley is considered

$$Ej^{(i)} = \langle d\varepsilon/dt \rangle^{(i)} \tag{64}$$

where $j^{(i)} = e n^{(i)} \tilde{\mu}^{(i)} E$, $\tilde{\mu}^{(i)}$ being the mobility tensor depending on T_e, and $j^{(i)}$ denotes the component of current due to valley i. $n^{(i)}$ is the number of electrons in valley i. The energy loss rate $\langle d\varepsilon/dt \rangle^{(i)}$ is the sum of contributions due to intra- (e.g. acoustic and optical phonon scattering) and intervalley loss rates. The contribution due to intervalley scattering is given by [193]

$$\langle d\varepsilon/dt \rangle_{iv}^{(i)} = \sum_{j \neq i} \langle d\varepsilon/dt \rangle_{l}^{i \to j} - \sum_{j \neq i} \langle d\varepsilon/dt \rangle_{g}^{j \to i}$$

where the first term describes the loss due to electrons scattered out of valley i and the second the gain due to electrons scattered from all other valleys into valley i. The loss is given by

$$\langle d\varepsilon/dt \rangle_{l}^{i \to j} = \langle \varepsilon/\tau_{iv}^{i \to j} \rangle^{(i)}$$

and the gain by

$$\langle d\varepsilon/dt \rangle_{g}^{j \to i} = (n^{(j)}/n^{(i)}) \langle (\varepsilon - \hbar\omega_{ji})/\tau_{iv}^{e,j \to i} + (\varepsilon + \hbar\omega_{ji})/\tau_{iv}^{a,j \to i} \rangle .$$

Here $\tau_{iv}^{i \to j}$ is the intervalley relaxation time for scattering from valley $i \to j$ [194] and $\tau_{iv}^{e,j \to i}$ is the relaxation time for scattering $(j \to i)$ due to emission of a phonon $\hbar\omega_{ij}$ and $\tau_{iv}^{a,j \to i}$ the analogous for absorption of $\hbar\omega_{ij}$.

In order to derive the number of electrons in the valleys the following equation is used

$$\sum_{j \neq i} (\partial n/\partial t)_{j \to i} = \sum_{j \neq i} (\partial n/\partial t)_{i \to j} \tag{65}$$

which is a balance equation for the electrons scattered in and out. The rate of transfer of carriers from valley i to valley j is given by

$$(\partial n/\partial t)_{i \to j} = - \langle 1/\tau_{iv}^{i \to j} \rangle n^{(i)} .$$

All brackets denote averages over the MB distribution with T_e as parameter [26].

[26] In performing the calculations the ellipsoidal surfaces of constant energy are usually transformed into spheres. See [192].

In Section 2.2.8 the comparison between experiment and theory for n-GaSb was simplified by assuming that the carrier temperature in the Γ valley and in the L valleys was the same. The energy loss rate was not calculated, but an experimental value for τ_ε was used [102, 104]. Then

$$eE^2 \left[\mu_\Gamma n_\Gamma(T_e) + \mu_L n_L(T_e)\right]/n = (\langle\varepsilon(T_e)\rangle - \varepsilon_L)/\tau_\varepsilon$$

$$\langle\varepsilon(T_e)\rangle = (4\pi k_B T_e/n)(2m_\Gamma k_B T_e/h^2)^{3/2}$$

$$\cdot\left[F_{3/2}(\varepsilon_F/k_B T_e) + N_{\Gamma,L} F_{3/2}((\varepsilon_F - \Delta_{\Gamma,L})/k_B T_e)\right] + \Delta_{\Gamma,L} n_L/n$$

$n = n_\Gamma(T_e) + n_L(T_e)$, $N_{\Gamma,L}$ is the density of states ratio for L valleys compared to the Γ valley and $\Delta_{\Gamma,L}$ the energetic separation between Γ and L.

The data on the hot electron Faraday effect given in Fig. 21 were also calculated using this model and in both cases satisfying agreement with the observed $T_e(E)$ and $n_\Gamma(E)/n_\Gamma(0)$ data was obtained [112, 195]. In a Monte Carlo calculation [196], in which ionized impurity scattering was included, even a better agreement with experiment was found. Faraday effect measurements in n-Ge [113] yielding information on equivalent and non equivalent intervalley scattering were interpreted with the method of energy balance just described [Eq. (64), (65)] but, the effect of the magnetic field was included [27].

Measurements of the birefrigence [116] (see Section 2.2.9) yielded beside the increase of T_e (Fig. 22) also information on the number of carriers in the particular valleys. Direct evidence for the equivalent intervalley scattering was obtained. In Fig. 46 the ratio of the number of carriers in the cold (n_c) to the hot (n_h) valleys is shown $(E \parallel [111])$ for n-Ge. Also included are data deduced from current voltage characteristics and obtained from the balance equations [116, 198–200]. For high carrier concentrations, the difference between n_c and n_h is reduced as shown in Fig. 46 which also resulted from calculations using the balance method [116].

Calculations based on the displaced MB distribution without making the diffusion approximation, where formerly applied to polar semiconductors (e.g. [201, 202]). High field $j - (E)$ characteristics (up to several kV/cm) were calculated at temperatures where already the momentum relaxation is dominated by polar optical scattering. However there is now a tendency to use for strong anisotropic and inelastic scattering mechanisms the more exact numerical methods [203] (see Sections 3.1.1, 3.1.2).

---

[27] In a many valley semiconductor a magnetic field will cause a change of the energy gain in each valley $Ej^{(i)}$ due to the altering of the electric field by the Hall field. Thus the repopulation of the electrons among the different valleys is directly influenced. Interesting phenomena like negative magnetoresistance can occur (Ref. [197]).

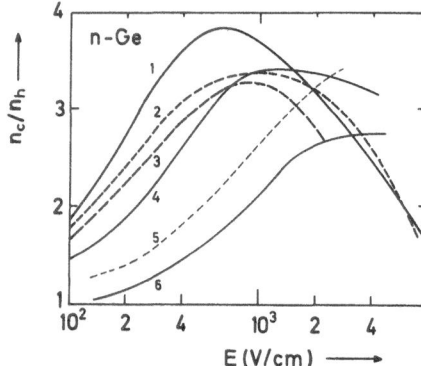

Fig. 46. Ratio of number of electrons in the cold [111] to the hot $\langle \bar{1}11 \rangle$ valley in n-Ge, $E \parallel [111]$. (1) Calculated balance equations ($T = 78$ K), Ref. [198]. (2) Deduced from mobility measurements ($T = 77$ K, $n \approx 1 \times 10^{14}$ cm^{-3}), Ref. [199]. (3) Microwave measurements ($T = 78$ K), Ref. [198]. (4) Experimental data from birefrigence ($T = 85$ K, $n = 5 \times 10^{14}$ cm^{-3}), Ref. [116]. (5) Deduced from mobility measurements ($T = 85$ K, $n = 3 \times 10^{14}$ cm^{-3}), Ref. [200]. (6) As (4) ($n = 1.6 \times 10^{15}$ cm^{-3}), Ref. [116]. Curves (4) and (6) can be reproduced approximately by balance calculations (Ref. [116]). Influence of increasing carrier concentration on the redistribution of the carriers is demonstrated. Results of balance equations depend on the choice of D, D_{ij}

Despite being an approximate model only, the electron temperature concept has prooved to yield useful results if it is applied under conditions where the energy gained from the field is randomized and the inelasticity of the scattering mechanisms is not pronounced. The calculations based on this model are relatively simple and physical insight on the dominant scattering mechanisms is rapidly gained. Often even complex phenomena like high field conduction in magnetic fields can be at least explained. This concept is still used under conditions where the electron-electron scattering is supposed to be of importance, since numerical solutions of the BE including this particular scattering mechanism are scarcely available.

4.2. Calculated Distribution Functions

The comparison of experimental data with calculated distribution functions is essentially based on the methods described in Section 3.1.1 and 3.1.2. There has been a considerable progress in numerical solutions of the BE: e.g. ionized impurity scattering [204], a generalization to ellipsoidal energy surfaces [205], a.c. conductivity [121, 206] can be treated and an increase in computing efficiency was achieved [121, 207]. Unfortunately this progress did not have an immediate influence on the experimental determination of distribution functions.

In p-Ge both Monte Carlo and path variable methods have been used for comparison with experimental data obtained from the electric field dependence of intervalence band absorption as shown in Figs. 29 and 30. The increase of the average energy with field, and v_d/v_{rms} are in good agreement with experiment for both methods. The distribution functions for 800 V/cm are also in satisfactory agreement with each other and with experiment. Differences occur only near the kink of the distribution function at the optical phonon energy which is in the experiment not so pronounced (Fig. 29). Budd [119] has attributed these deviations from experiment and those regarding the anisotropy of the distribution function (Fig. 32), which are also present in the Monte Carlo calculations, to the neglection of carrier-carrier scattering.

This example demonstrates that a sufficient description of the experimental data is still not possible. The mean values, which are averaged over the whole distribution are well represented. However if the detailed structure of the distribution function is compared with experiment deviations show up. Therefore an improvement of the calculations, by the suggested inclusion of carrier-carrier scattering would be desirable[28].

A lot of theoretical data have been published on the shape of the hot electron distribution function in n-GaAs (e.g. [203]). Experimental data on the distribution function in n-GaAs at 77 K obtained from measuring the recombination radiation spectrum were compared with iterative calculations by Rees [121] and with a displaced MB distribution ($v_d = 2.41 \times 10^7$ cm s^{-1}) in Fig. 39. The experimental method is however until now not sensitive enough to distinguish between different theoretical models. Complications due to inhomogeneities in the samples are the reason for the failure.

In the experiment by Mooradian and McWorther [171] on inelastic light scattering by hot electrons no direct information on the distribution function was given. It was however shown that at $E = 1400$ V/cm and $T \approx 25$ K a drifted Maxwellian does not account for the observed light scattering spectrum (Fig. 44). In principle it should be possible to adjust a distribution function such as to yield the observed light scattering spectrum. Unfortunately this has however not been made.

A comparison of the distribution functions derived for degenerate n-GaAs at 77 K at $E = 200$, 300, and 450 V/cm with theory can until now not be performed (Fig. 41). There are no calculations available which apply for degenerate statistics. The method of the field modulation of the fundamental absorption yields the detailed form of the distribution function rather than a functional only. The experiments have also shown

---

[28] Persky and Bartelink [208] have used a parametrized model of the distribution function, where the energy dependence is represented by 2 MB distributions at different effective temperatures below and above the optical phonon energy. No assumptions on the angular dependence of the distribution function were made. Their results are approximately the same as those of Budd [119].

definitely that despite the high carrier concentrations, deviations from a Fermi distribution exist which are outside the experimental error.

Since it was pointed out [11, 121] that in principle degeneracy can be included in iterative calculations, it should be worth-while to make the computations.

The experimental determination of distribution functions is difficult since it is necessary to use optical methods together with fast pulse techniques. The number of different methods is small (until now four) and some are restricted to particular band shapes. In addition only in *p*-Ge several authors have experimentally derived distribution functions whereas the other methods were only tried once. Thus, except for *p*-Ge, there is no possibility for a comparison of different data. In all investigations simplifying assumptions had to be made. For some of these assumptions it is difficult to estimate their influence on the final shape of the derived distribution function (see Section 3.2.2).

There is now no question about the importance of detailed information on the shape of the distribution function, since it directly reflects the important scattering mechanisms (see Figs. 23, 25, and 29) rather than some average over the distribution function only. Since at least in principle distribution functions can be calculated for a variety of scattering mechanisms more work should be devoted to this subject in order to check the various theoretical models and to improve them.

5. Summary

In this paper we have summarized the experimental methods which have been used to get information on the increase of mean carrier energy characterized by the electron temperature or the change of the carrier distribution function in high electric fields. For the determination of mean values, characterizing the microscopic changes induced by high fields, transport and optical measurements can be employed. However detailed information on the shape of the carrier distribution functions can only be obtained from optical experiments.

Although the term electron temperature already implies an important assumption on the shape of the distribution function made in the analysis of a hot electron experiment, this concept is still widely used. Experimental determinations of electron temperatures have contributed a lot to a better understanding of the physical principles of the interaction of electrons and phonons. New energy loss processes like the emission of pairs of transverse acoustic phonons at low lattice temperatures have been found. The deformation potential constant for several III–V compounds was deduced from experiments at low temperatures.

The time dependent increase of the mean carrier energy after application of high fields was directly observed in nondegenerate as well as degenerate semiconductors. Optical measurements like the Faraday effect and birefrigence have yielded information on the repopulation of the carriers among different conduction band valleys. The modulation of fundamental absorption by high fields was exploited to deduce electron temperatures in degenerate semiconductors.

A refinement in the analysis of hot electron transport properties was achieved through Monte Carlo and iterative calculations which can be performed for complex band structures and scattering mechanisms. However the experimental methods for determining distribution functions are still restricted to few semiconductors and a comparison of experimental and numerical results has revealed that either the theoretical models have still to be improved (p-Ge) or that the experimental resolution is still insufficient to distinguish between simple Maxwell-Boltzmann distributions and exactly deduced distributions. There are indications that carrier-carrier scattering plays an important role on determing the shape of the hot carrier distribution functions even at relatively small carrier concentrations. However this particular scattering mechanism has scarcely been incorporated in the numerical solutions of the Boltzmann transport equation.

Intervalence band absorption in p-Ge has yielded information on the distribution functions of heavy and light holes which exhibit considerable differences from a simple Maxwell-Boltzmann distribution. They can be approximated by a two temperature Maxwell-Boltzmann distribution with different effective temperatures below and above the optical phonon energy. The detection of electron-hole recombination radiation, inelastic light scattering and change of the fundamental absorption edge has been exploited to deduce hot electron distribution functions in n-GaAs.

The progress achieved in the numerical solution of the Boltzmann equation has been possible due to the availability of powerful computers. The experimental improvements are based on new pulse techniques, better electronic instrumentation, and the use of lasers. Also new crystal growing technologies have provided high quality pure single crystals. There are still interesting problems to be solved concerning the experimental investigation of distribution functions. Recently [209] it was predicted that a population inversion of hot carrier distribution function in k-space may occur under crossed electric and magnetic fields and analogies to the energy levels of a laser have been put forward. Population inversion also occurs as a combination of the effect of polar optical and intervalley scattering in GaAs. None of these effects has been verified until now by an experimental probing of the distribution function.

Acknowledgments. I want to express my appreciation to Prof. P. Grosse for initiating this work, and for many helpful discussions during the preparation of the manuscript. I am indebted to Dr. G. Nimtz and Prof. H. Heinrich for a critical reading of the manuscript and stimulating discussions. I wish to thank Dr. W. Jantsch and Prof. H. Heinrich for providing their results prior to publication, and Drs. K. Hess, P. D. Southgate, E. G. S. Paige, O. Christensen for helpful correspondence on certain topics of this paper, and Dipl. Phys. U. Kaiser-Dieckhoff for translating russian references.

Notes Added in Proof. In the meantime a number of further investigations have been performed which are listed below. Reference is made to each chapter.

2.2.3: FIR cyclotron resonance measurements in n-InSb have yielded values of T_e and occupation numbers of carriers in different Landau-levels as a function of E (Kobayashi, K. L. I., Otsuka, E.: J. Phys. Chem. Solids **35**, 839 (1974)).

In microwave cyclotron resonance experiments a selective heating of holes in p-Ge has been achieved, thus getting information on scattering phenomena including interband transitions (Gershenzon, E. M., Gusinskii, E. N., Rabinovich, R. I., Soina, N. V.: Phys. Stat. Sol. (b) **64**, 367 (1974)). Using another $g(\varepsilon)$ dependence, the measurements described in [63] have been reevaluated by Zawadzki, W.: In: Wallace, P. R., Harris, R., Zuckermann, M. J. (Eds.): New developments in semiconductors, p. 441. Leyden: Noordhoff International Publishing 1973.

2.2.9: Hot-electron Faraday effect measurements have been performed in the millimeter wavelength range using microwave heating in n-InSb (Potapov, V. T., Sokolovskii, A. V., Trifonov, V. I., Yaremeko, N. G.: Sov. Phys. Semiconductors **6**, 1081 (1973)) following earlier theoretical suggestions (Chakravarti, A. K.: J. Appl. Phys. **42**, 2875 (1971); Sodha, M. S., Sharma, S. K., Dubey, P. K.: J. Appl. Phys. **42**, 2400 (1971)). The anisotropy of the refractive index in strong E in n-Si has been exploited for a determination of n_c, n_h, and T_c, T_h (Vorob'ev, L. E., Stafeev, V. I., Ushakov, A. Yu.: Sov. Phys. Semiconductors **7**, 624 (1973)).

3.2.3: The recombination radiation after impact ionization of Sb-doped Ge has been analyzed and carrier distribution functions have been determined for ε up to 8 meV at $T \approx 4\,\mathrm{K}$ (Thomas, S. R., Fan, H. Y.: Phys. Rev. B **9**, 4295 (1974)).

Ulbrich, R. (Phys. Rev. B **8**, 5719 (1973)) has determined the dynamic change of an electron distribution function using n-GaAs at $T = 2-4$ K after photoexcitation within a resolution of 0.2 ns up to energies of 15 meV. Additional investigations on carrier heating by photoexcitation were made by Shah, J.: Phys. Rev. B **9**, 562 (1974).

4.4.2: A quantum transport theory of high field conduction was given by Barker, J. R.: J. Phys. C (Solid State Physics) **6**, 2663 (1973). A new main field in the study of T_e and distribution functions is concerned with transport in two-dimensional systems: Hess, K., Sah, C. T.: J. Appl. Phys. **45**, 1254 (1974).

List of Symbols

a	polarization vector
A	Shubnikov-de Haas amplitudes
b_s	oscillatory component in SdH effect
B	magnetic field strength
B_d	arrival probability
c	velocity of light
c_h	specific heat
d	mass density
d'	sample thickness

D	optical phonon coupling constant
D_{ij}	intervalley coupling constant
e	electronic charge
e_{14}	piezoelectric coupling constant
E	electric field strength
E_1	acoustic deformation potential constant
$f(k), f(\varepsilon)$	distribution function
f_0, f_1, f_2	coefficients in a Legendre polynomials expansion
F_k	Fermi integrals
g	gyromagnetic ratio, g-factor
g_{ci}	oscillator strength
H'	perturbing potential
j	current density
k	wave vector of a carrier
k_B	Boltzmann constant
k_d	wave vector corresponding to drift velocity v_d
k_F	Fermi wave vector
K	absorption constant
K_E	absorption constant in an electric field
K_{th}	absorption constant in thermal equilibrium
l_e	mean free path
m	effective mass
m_0	free electron mass
m_l, m_t	longitudinal and transverse mass in an ellipsoidal band model
$M(\alpha_c, \alpha_i)$	combined mass factor
N_I	total impurity density
n	carrier concentration
n_c, n_h	carrier concentration in a cold and hot valley
n_r	refractive index
n_Γ, n_L	carrier concentration in Γ valley, L valleys
N	Landau level number
$N_c, N_v, N_{\Gamma,L}$	density of states factors
N_q	phonon occupation number
O_i	occupancy
p	hole concentration
P	energy loss rate
P_D	probability distribution function
$P_l(\varepsilon)$	Luminescence spectrum
P_n	Legendre polynomials
q	phonon wave vector
r	vector in space
r	Hall coefficient factor
$r_{0,I}$	Hall coefficient factor due to ionized impurity scattering
R_B	Hall constant
Ry	Rydberg energy
s	spin, quantum number, $S(\varepsilon)$ spectral sensitivity
t	time
T	lattice temperature
T_e	electron temperature
T_D	Dingle temperature
T_F	Fermi temperature
$T_n, T_{n'}$	noise temperature

$T(\varepsilon)$	transmittivity
T_1, T_2	effective temperature of MB distribution below and above $\hbar\omega_0$
u	longitudinal sound velocity
v	carrier velocity
v_d	drift velocity
v_{rms}	root mean square velocity
V	volume
$W(k, k')$	transition probability
α_i	m_0/m_i mass ratios
Γ	Γ-function
Δ	spin orbit splitting
$\Delta\varepsilon_g$	gap energy
$\Delta_{\Gamma, L}$	energetic separation of Γ and L valleys
ε	carrier energy
$\langle\varepsilon\rangle$	mean carrier energy
ε_B	Burstein shift
$\varepsilon_F, \varepsilon_{F_0}$	Fermienergy, Fermienergy at $T=0$
ε_L	carrier energy in thermal equilibrium
η	reduced Fermienergy
ϑ	Faraday rotation angle
θ	$\sphericalangle(k, E)$
θ_0	optical phonon Debye temperature
$\varkappa$	static dielectric constant
$\varkappa_\infty$	optical dielectric constant
λ	scattering rate
λ_D	screening length
μ	mobility
μ_0	ohmic mobility
$\mu_{0,I}$	mobility due to ionized impurity scattering
ν	ratio of spin splitting to Landau level spacing
$\tilde{\nu}$	wave number
Ξ_u, Ξ_d	deformation potential constants for ellipsoidal energy surfaces
ϱ	resistivity
ϱ_{ij}	component of magnetoresistance tensor
σ	conductivity
σ_{ij}	component of conductivity tensor
$\Delta\sigma$	absorption cross section (K/p)
$\mathrm{d}^2\sigma/\mathrm{d}\Omega\,\mathrm{d}\omega$	scattering cross section
τ	mean free time, relaxation time
τ_a	mean free time due to acoustic phonon scattering
τ_ε	energy relaxation time
τ_m	momentum relaxation time
Φ	constant characterizing self scattering
ω	angular frequency
ω_0	angular frequency of optical phonon
ω_c	cyclotron resonance frequency
ω_q	angular frequency of phonon with wave vector q
ω_1, ω_2	incident and scattered light frequencies

References

1. Landau, L., Kompanejez, A.: Phys.Z.S.U. **6**, 163 (1934). Davydov, B.Ch., Smushkevich, I.M.: Z. Eksper. i Teor. Fiz. **7**, 1069 (1937). — Phys. Z.S.U. **12**, 269 (1937).
2. Shockley, W.: Bell Syst. Tech. J. **30**, 990 (1951).
3. Gunn, J.B.: In: Gibson, A.F. (Ed.): Progress in semiconductors, Vol. 2, p. 213. New York: J. Wiley 1957.
4. Paige, E.G.S.: In: Gibson, A.F. (Ed.): Progress in semiconductors, Vol. 8, p. 1. New York: J. Wiley 1964.
5. Reik, H.G.: In: Sauter, F. (Ed.): Festkörperprobleme, Vol. 1, p. 89. Braunschweig: . F. Vieweg und Sohn 1962.
6. Schmidt-Tiedemann, K.J.: In: Sauter, F. (Ed.): Festkörperprobleme, Vol. 1, p. 122. Braunschweig: F. Vieweg und Sohn 1962.
7. Nag, B.R.: Solid State Electr. **10**, 385 (1967); Theory of electrical transport in semiconductors. Oxford: Pergamon Press Ltd. 1972.
8. Asche, M., Sarbei, O.G.: Phys. Stat. Sol. **33**, 9 (1969).
9. Dienys, V., Pozhela, Yu.K.: Hot electrons. Vilnius: Lithuanian Academy of Sciences 1971.
10. Conwell, E.M.: In: Seitz, F., Turnbull, D., Ehrenreich, H. (Eds.): Solid state physics, Suppl. 9. New York: Academic Press Inc. 1967.
11. Fawcett, W.: Proc. Int. Conf. Physics of Semiconductors, Cambridge Mass. Keller, S.P., Hensel, J.C., Stern, F. (Eds.). Oak Ridge, Tenn.: USAEC, 1970, p. 51; Non-Ohmic Transport in Semiconductors, in: Salam, A. (Ed.): Electrons in Crystalline Solids, p. 531. Vienna: IAEA, 1973.
12. Kubo, R.: In: Cohen, M.H., Thirring, W. (Eds.): 100 Years Boltzmann equation, p. 301. Wien: Springer 1973.
13. Smirnow, W.I.: Lehrgang der Höheren Mathematik, Vol. III/2, p. 411. Berlin: VEB Deutscher Verlag der Wissenschaften 1964.
14. Irving, J., Mullineux, N.: Mathematics in physics and engineering, p. 194. New York: Academic Press 1959.
15. Fröhlich, H., Paranjape, B.V.: Proc. Phys. Soc. (London) B **69**, 21 (1956).
16. Pines, D.: Phys. Rev. **92**, 626 (1953).
17. Hearn, C.J.: Proc. Phys. Soc. (London) **86**, 881 (1965).
18. Hasegawa, A., Yamashita, J.: J. Phys. Soc. Japan **17**, 1751 (1962).
19. Dykman, I.M., Tomchuk, P.M.: Sov. Phys. Solid State **2**, 1988 (1961). — Sov. Phys. Solid State **3**, 1393 (1962).
20. Stratton, R.: Proc. Roy. Soc. (London) A **242**, 355 (1957). — Proc. Roy. Soc. (London) A **246**, 406 (1958).
21. Spitzer, L.: Physics of fully ionized gases. New York: Inter-Science Publishers, 1967, Appendix.
22. Hess, K.: Private communication. — Seeger, K., Pötzl, H.: See Ref. [12], p. 341. — Seeger, K.: Semiconductor physics, p. 57. Wien: Springer 1973
23. Blakemore, J.S.: Semiconductor statistics. Oxford-London-New York-Paris: Pergamon Press 1962.
24. Katayama, Y., Tanaka, S.: Phys. Rev. **153**, 873 (1967).
25. Miyazawa, H.: J. Phys. Soc. Japan **26**, 700 (1969).
26. Maneval, J.P., Zylbersztejn, A., Budd, H.F.: Phys. Rev. Letters **23**, 848 (1969).
27. Szymanska, W., Maneval, J.P.: Solid State Commun. **8**, 879 (1970).
28. Martin, J.P., Mead, J.B.: Appl. Phys. Letters **17**, 320 (1970).
29. Kinch, M.A.: Brit. J. Appl. Phys. **17**, 1257 (1966); Proc. Phys. Soc. (London) **90**, 819 (1967).

30. Zylbersztejn,A.: Proc. Int. Conf. Physics of Semiconductors, p. 505. Hulin,M. (Ed.). Paris: Dunod 1964.
31. Sandercock,J.R.: Proc. Phys. Soc. (London) **86**, 1221 (1965).
32. Sandercock,J.R.: Solid State Commun. **7**, 721 (1969).
33. Lifshits,T.M., Oleinikov,A.Ya., Shulman,A.Ya.: Phys. Stat. Sol. **14**, 511 (1966).
34. Peskett,G.D., Rollin,B.V.: Proc. Phys. Soc. (London) **82**, 467 (1963).
35. Sladek,R.J.: Phys. Rev. **120**, 1589 (1960).
36. Whalen,J.J., Westgate,C.R.: Appl. Phys. Letters **15**, 292 (1969).
37. Whalen,J.J., Westgate,C.R.: J. Appl. Phys. **43**, 1965 (1972).
38. Racek,W., Bauer,G.: Verh. DPG (VI) **8**, 331 (1973).
39. Yafet,Y., Keyes,R.W., Adams,E.N.: J. Phys. Chem. Solids **1**, 137 (1956).
40. e.g. Ortenberg,M. von: J. Phys. Chem. Solids **34**, 397 (1973).
41. Crandall,R.S.: Solid State Commun. **7**, 1575 (1969).
42. Dyakonov,M.I., Efros,A.L., Mitchell,D.L.: Phys. Rev. **180**, 813 (1969).
43. Kaufman,L.A., Neuringer,L.J.: Phys. Rev. B**2**, 1840 (1970).
44. Beer,A.C.: In: Seitz,F., Turnbull,D. (Eds.): Galvanomagnetic effects in semiconductors, Suppl. 4. New York: Academic Press 1963
45. Bauer,G., Kahlert,H.: Phys. Rev. B**5**, 566 (1972).
46. Budd,H.F.: Phys. Rev. **131**, 1520 (1963). — Phys. Rev. **140**, A2170 (1965).
47. Matz,D., Garcia-Molinar,F.: Phys. Stat. Sol. **5**, 495 (1964).
48. Crandall,R.S.: Phys. Rev. **169**, 585 (1968).
49. e.g. Ziman,J.M.: Principles of the theory of solids. Cambridge: University Press 1964.
50. Kubo,R., Miyake,S.J., Hashitsume,W.: In: Seitz,F., Turnbull,D., Ehrenreich,H. (Eds.): Solid state physics, Vol. 17, p. 269. New York: Academic Press 1965.
51. Miyazawa,H., Ikoma,H.: Solid State Commun. **5**, 229 (1967). — J. Phys. Soc. Japan **23**, 290 (1967).
52. Kobayashi,S.: Proc. Symp. Submillimeter Waves, p. 331. Fox,J. (Ed.). New York: Polytechnic Press 1970.
    Kobayashi,S., Watanabe,M.: Proc. Int. Conf. Physics of Semiconductors, p. 909. Miasek,M. (Ed.). Warsaw: Polish Scientific Publishers 1972.
53. Crandall,R.S.: Phys. Rev. B**1**, 730 (1970).
54. Crandall,R.S.: J. Phys. Chem. Solids **31**, 771 (1970).
55. Kazarinov,R.F., Skobov,V.G.: Sov. Phys. JETP **15**, 726 (1962).
    Zlobin,A.M., Zyryanov,P.S.: Sov. Phys. Uskekhi **14**, 379 (1972).
56. Fujisada,H., Kataoka,S., Beer,A.C.: Phys. Rev. B**3**, 3249 (1971).
57. Kotera,N., Komatsubara,K.F., Yamada,E.: J. Phys. Soc. Japan Suppl. **21**, 411 (1966). — J. Phys. Chem. Solids **33**, 1311 (1972).
58. Yamada,E., Kurosawa,T.: Proc. Int. Conf. Physics of Semiconductors, Moscow, p. 805. Ryvkin,S. (Ed.). Leningrad: Publishing House Nauka 1968.
59. Lax,B., Mavroides,J.G., Zeiger,H.J., Keyes,R.J.: Phys. Rev. **122**, 31 (1961).
60. Isaacson,R.A.: Phys. Rev. **169**, 312 (1968).
    Kaplan,D., Konopka,J.: See Ref. [58], p. 1146.
61. Isaacson,R.A., Feher,G.: Bull. Am. Phys. Soc. **7**, 484 (1962).
62. Guéron,M.: See Ref. [30], p. 433.
63. Konopka,J.: Phys. Rev. Letters **24**, 666 (1970).
64. Roth,L.M., Lax,B., Zwerdling,S.: Phys. Rev. **114**, 90 (1959).
65. Roth, L.M., Argyres,P.N.: In: Willardson,R.K., Beer,A.C. (Eds.). Semiconductors and semimetals, Vol. 1, p. 159. New York. Academic Press 1966.
66. Dingle,R.B.: Proc. Phys. Soc. (London) A**211**, 517 (1952).
67. Komatsubara,K.F.: Phys. Rev. Letters **16**, 1044 (1966).
68. Isaacson,R.A., Bridges,F.: Solid State Commun. **4**, 635 (1966).
69. Konopka,J.: Solid State Commun. **5**, 809 (1967).

70. Bykovskii, Yu. A., Elesin, V. F., Kadushkin, V. I., Protasov, E. A.: JETP Letters **10**, 149 (1969).
71. Neuberger, M.: Handbook of electronic materials, Vol. 2, p. 93. New York: IFI Plenum Press 1971.
72. Bauer, G.: Verhandl. DPG **5**, 324 (1970).
73. Bauer, G., Kahlert, H.: See Ref. [11], p. 65.
74. Bauer, G., Kahlert, H.: Phys. Rev. B**5**, 566 (1972).
75. Kahlert, H., Bauer, G.: Phys. Stat. Sol. (b) **46**, 535 (1971).
76. Kahlert, H., Bauer, G.: Phys. Rev. B**7**, 2670 (1973).
    Bauer, G., Kahlert, H.: J. Phys. C (Solid State Physics) **6**, 1253 (1973).
77. Bauer, G., Kahlert, H.: Phys. Letters **41** A, 351 (1972).
78. Bauer, G., Kahlert, H., Kocevar, P.: Int. Symp. High Field Transport in Semiconductors. University of Modena, 1973 (unpublished).
79. Harper, P. G., Hodby, J. W., Stradling, R. A.: Rep. Prog. Physics **36**, 1 (1973).
80. Wilson, A. H.: Theory of metals, p. 147. Cambridge: University Press 1965.
81. Bauer, G.: (Unpublished).
82. Erlbach, E., Gunn, J. B.: Phys. Rev. Letters **8**, 280 (1962).
83. Erlbach, E., Gunn, J. B.: Proc. Int. Conf. Physics of Semiconductors, Exeter, p. 128. Stickland, A. C. (Ed.). London: Institute of Physics and Physical Society 1962.
84. Price, P. J.: IBM J. Res. Develop. **3**, 191 (1959).
85. Price, P. J.: In: Burgess, R. E. (Ed.): Fluctuation phenomena in solids, p. 355. New York: Academic Press 1965.
86. Hart, L. G.: Can. J. Physics **48**, 531 (1970); **49**, 1469 (1971).
87. Bareikis, V., Pozhela, Yu. K., Matulioniene, I.: See Ref. [58], p. 760.
    Bareikis, V., Shaltis, R., Pozhela, Yu. K.: Lit. Fiz. Rinkinys **6**, 99 (1966); **6**, 437 (1966). See Ref. [9], p. 219.
88. Nougier, J. P.: Physica **64**, 209 (1973).
    Nougier, J. P., Rolland, M.: Phys. Rev. B **8**, 5728 (1973); Sol. State Electron. **16**, 1399 (1973).
88a. Nag, B. R., Robson, P. N.: Phys. Letters A **43**, 507 (1973).
89. Smith, R. A.: Semiconductors, p. 165. Cambridge: University Press 1959.
90. Ref. [10], p. 45, p. 276.
91. Stenbek, M.: Izv. Akad. Nauk SSSR **22**, 1560 (1956).
92. Bok, J.: Solid State Phys. Electron. Telecommun. Proc. Int. Conf. Brussels: 1958, Vol. 1, p. 475. New York: Academic Press 1960.
93. Pozhela, Yu. K., Repshas, K. K., Shilalnikas, V. I.: Ref. [83], p. 149.
94. Zucker, J.: J. Appl. Phys. **35**, 618 (1964).
95. Conwell, E. M., Zucker, J.: J. Appl. Phys. **36**, 2192 (1965).
96. Hamaguchi, C., Inuishi, Y.: J. Phys. Chem. Solids **27**, 1511 (1966).
97a. Alekseenko, M. V., Veinger, A. I.: Sov. Phys. Semiconductors **5**, 1952 (1972).
97b. Sharma, S. K., Dubey, P. K.: J. Appl. Phys. **42**, 2512 (1971).
97. See Ref. [9], p. 167.
98. Burstein, E.: Phys. Rev. **93**, 632 (1954).
99. Shur, M. S.: Phys. Letters **29** A, 490 (1969).
100. Franz, W.: Z. Naturforsch. **13a**, 484 (1958).
101. Keldysh, L. V.: Sov. Phys. JETP **7**, 788 (1958).
102. Heinrich, H., Jantsch, W.: Phys. Rev. B**4**, 2504 (1971).
103. Jantsch, W.: PhD thesis, University of Vienna, 1971 (unpublished).
     Seeger, K., Bauer, G., Kuchar, F., Kuzmany, H.: Acta Phys. Austriaca **35**, 195 (1972).
104. Heinrich, H., Hess, K., Jantsch, W., Pfeiler, W.: J. Phys. Chem. Solids **33**, 425 (1972).
105. Balkanski, M., Amzallag, E.: Phys. Stat. Sol. **30**, 407 (1968).
106. e.g. Rau, R. R., Caspari, M. E.: Phys. Rev. **100**, 632 (1955).

107. Subashiev, A. V.: Sov. Phys. Solid State **7**, 751 (1965).
Gulyaev, Yu. V.: JETP Letters **1**, 81 (1965).

108. Vorob'ev, L. E., Stafeev, V. I., Ushakov, A. U., Shturbin, A. V.: Ref. [58], p. 765; Sov. Phys. Semiconductors **1**, 114 (1967).

109. Wood, V. E.: J. Appl. Phys. **40**, 3740 (1969).
Chattopadhyay, D., Nag, B. R.: Phys. Stat. Sol. **40**, 701 (1970).

110. Almazov, L. A.: Phys. Stat. Sol. (b) **58**, 821 (1973) and references cited.

111. Vorob'ev, L. E., Komissarow, V. S., Stafeev, V. I.: Sov. Phys. Semiconductors **7**, 59 (1973); Phys. Stat. Sol. (b) **52**, 25 (1972); Phys. Stat. Sol. (b) **54**, K 61 (1972).

112. Heinrich, H.: Phys. Letters **32** A, 331 (1970); Phys. Rev. B **3**, 416 (1971).
Asche, M.: Phys. Stat. Sol. **41**, 67 (1970).

113. Kriechbaum, M., Lischka, K., Kuchar, F., Heinrich, H.: Proc. Int. Conf. Physics of Semiconductors, p. 615. Miasek, M. (Ed.). Warszawa: Polish Scientific Publishers 1972.
Lischka, K., Heinrich, H.: Solid State Commun. **12**, 1187 (1973).

114. Ipatova, I. P., Kazarinov, R. F., Subashiev, A. V.: Sov. Phys. Solid State **7**, 1714 (1966).

115. Schmidt-Tiedemann, K. J.: Phys. Rev. Letters **7**, 732 (1961), see also Ref. [6], p. 122.

116. Vorob'ev, L. E., Stafeev, V. I., Ushakov, A. V.: Phys. Stat. Sol. (b) **53**, 431 (1972).

117. Reik, H. G., Risken, H.: Phys. Rev. **124**, 777 (1961).

118. Kurosawa, T.: J. Phys. Soc. Japan Suppl. **21**, 424 (1966); **31**, 668 (1971).

119. Budd, H. F.: J. Phys. Soc. Japan Suppl. **21**, 420 (1966); Phys. Rev. **158**, 798 (1967).

120. Fawcett, W., Boardman, A. D., Swain, S.: J. Phys. Chem. Solids **31**, 1963 (1970).

121. Rees, H. D.: J. Phys. Chem. Solids **30**, 643 (1969).
Rees, H. D.: IBM J. Res. Develop. **13**, 537 (1969).
Rees, H. D.: J. Phys. C (Solid State Phys.) **5**, 641 (1972).
Price, P. J.: Ref. [58], p. 753.

122. Kranzer, D., Hillbrand, H., Pötzl, H., Zimmerl, O.: Acta Phys. Austriaca **35**, 110 (1972).

123. Boardman, A. D., Fawcett, W., Ruch, J. G.: Phys. Stat. Sol. (a) **4**, 133 (1971).

124. Chou, S. C.: J. Appl. Phys. **43**, 1693 (1972).
Curby, R. C., Ferry, D. K.: Phys. Stat. Sol. (a) **15**, 319 (1973).

125. Lebwohl, P. A., Marcus, P. M.: Solid State Commun. **9**, 1971 (1971).
Bacchelli-Montefusco, L., Jacoboni, C.: Solid State Commun. **10**, 71 (1972).

126. Brown, M. A. C. S., Paige, E. G. S.: Phys. Rev. Letters **7**, 84 (1961).

127. Brown, M. A. C. S., Paige, E. G. S., Simcox, L. N.: Ref. [83], p. 111.

128. Pinson, W. E., Bray, R.: Bull. Am. Phys. Soc. **8**, 253 (1963).

129. Pinson, W. E., Bray, R.: Phys. Rev. **136**, A 1449 (1964).

130. Bray, R., Pinson, W. E.: Phys. Rev. Letters **11**, 502 (1963).

131. Baynham, A. C., Paige, E. G. S.: Phys. Letters **6**, 7 (1963).

132. Bray, R., Pinson, W. E., Brown, D. M., Kumar, C. S.: Ref. [30], p. 467.

133. Vasileva, M. A., Vorob'ev, L. E., Stafeev, V. I.: Sov. Phys. Semiconductors **1**, 273 (1967).
Vasileva, M. A., Vorob'ev, L. E., Stafeev, V. I.: Sov. Phys. Semiconductors **1**, 21 (1967).

134. Vasileva, M. A., Vorob'ev, L. E., Soltamov, U. B., Stafeev, V. I., Ushakov, A. Yu., Shturbin, A. V.: Sov. Phys. Semiconductors **1**, 361 (1967).

135. Baynham, A. C.: Solid State Commun. **3**, 253 (1965).

136. Christensen, O.: Phys. Rev. B **7**, 763 (1973); Ph.D. thesis, Lyngby 1971.

137. Briggs, H. B., Fletcher, R. G.: Phys. Rev. **91**, 1342 (1953).

138. Newman, R., Tyler, W. W.: Phys. Rev. **105**, 885 (1957).

139. Kahn, A. H.: Phys. Rev. **97**, 1647 (1955).

140. Kane, E. O.: J. Phys. Chem. Solids **1**, 82 (1956).

141. Fawcett, W.: Proc. Phys. Soc. (London) **85**, 931 (1965).

142. Arthur, J. B., Baynham, A. C., Fawcett, W., Paige, E. G. S.: Phys. Rev. **152**, 740 (1966).

143. Balslev, I.: Phys. Rev. 177, 1173 (1969).
144. Kessler, F. R., Kneser, Ch.: Phys. Stat. Sol. (b) 43, K 33 (1971).
145. Houghton, J. T., Smith, S. D.: Infrared Physics, p. 131. Oxford: Clarendon Press 1966.
146. Kessler, F. R.: Phys. Stat. Sol. 5, 3 (1964); Phys. Stat. Sol. 6, 3 (1964).
147. Baynham, A. C., Paige, E. G. S.: Ref. [30], p. 149.
148. McLean, T. P., Paige, E. G. S.: Ref. [83], p. 450.
149. Kessler, F. R., Kneser, Ch.: Z. Phys. 193, 266 (1966).
150. Vorob'ev, L. E., Stafeev, V. I.: Sov. Phys. Semiconductors 1, 1190 (1967).
151. Vorob'cv, L. E., Stafccv, V. I., Ushakov, A. Yu., Shturbin, A. V.: Ref. [58], p. 765.
152. Elliot, R. J.: Phys. Rev. 108, 1384 (1957).
153. Southgate, P. D.: Private communication.
154. Southgate, P. D., Hall, D. S., Dreeben, A. B.: J. Appl. Phys. 42, 2868 (1971).
155. Southgate, P. D., Hall, D. S.: Appl. Phys. Letters 16, 280 (1970).
156. Southgate, P. D.: Appl. Phys. Letters 15, 95 (1969).
157. Solomon, S. N.: PhD thesis, Purdue University, 1969 (unpublished).
158. Kaminskii, A. S., Pokrovskii, Ya. E., Svistunova, K. I.: Sov. Phys. Semiconductors 4, 1625 (1971).
159. Vasileva, M. A., Vorob'ev, L. E., Stafeev, V. I.: Sov. Phys. Semiconductors 3, 1144(1970).
160. Redfield, D., Afromowitz, M. A.: Appl. Phys. Letters 11, 138 (1967).
161. Rees, H. D.: Solid State Commun. 5, 365 (1967).
162. Dumke, W. P.: Phys. Rev. 108, 1419 (1957).
163. Jantsch, W., Heinrich, H.: Verhandl. DPG (VI) 8, 332 (1973) Solid State Commun. 13, 715 (1973).
164. Lambert, L. M.: J. Phys. Chem. Solids 26, 1409 (1965).
165. McGroddy, J. C., Christensen, O.: Bull. Am. Phys. Soc. Ser. II, 17, 325 (1972).
166. Hwang, C. J.: J. Appl. Phys. 41, 2668 (1970).
    Lefevre, J., Bois, D., Pinard, P., Davoine, F., Leclerc, P.: J. Opt. Soc. Am. 58, 1230 (1968).
167. Wolff, P. A.: In: Wright, G. B. (Ed.): Light scattering spectra of solids, p. 273. Berlin-Heidelberg-New York: Springer 1969.
168. Pines, D., Nozieres, P.: The theory of Fermi liquids. New York: W. A. Benjamin 1966.
169. Pines, D.: Elementary excitations in solids. New York: W. A. Benjamin 1963.
170. Hamilton, D. C., McWorther, A. L.: Ref. [167], p. 309.
171. Mooradian, A., McWorther, A. L.: Ref. [11], p. 380.
172. Mooradian, A.: Ref. [167], p. 285.
    Mooradian, A., McWorther, A. L.: Ref. [167], p. 297.
    Mooradian, A.: In: Madelung, O. (Ed.): Advances in solid state physics. Festkörperprobleme, IX, p. 73. Braunschweig: Vieweg 1969.
173. Mooradian, A., Foyt, A. G.: Bull. Am. Phys. Soc. 15, 303 (1970).
174. See Ref. [10], p. 164.
175. Ziman, J. M.: Principles of the theory of solids. Cambridge: University Press 1964.
176. See Ref. [44], p. 113.
177. Greene, R. F.: J. Electr. Control 3, 387 (1957).
178. Kogan, Sh. M.: Sov. Phys. Solid State 4, 1813 (1963).
179. Ehrenreich, H.: J. Phys. Chem. Solids 9, 129 (1959).
180. Ziman, J. M.: Electrons and phonons. Oxford: Clarendon, 1963.
181. Peierls, R.: Helv. Phys. Acta Suppl. 7, 24 (1934).
182. Peierls, R.: Quantum theory of solids. Oxford: University Press 1955.
183. Tsidilkovskii, I. M., Demchuk, K. M.: Phys. Stat. Sol. (b) 44, 293 (1971).
184. Stradling, R. A., Wood, R. A.: J. Phys. C (Solid State Physics) 3, 2425 (1970).
185. Racek, W., Bauer, G., Kahlert, H.: Phys. Rev. Letters 31, 301 (1973).
186. Spitzer, W. G.: In: Willardson, R. K., Beer, A. C. (Eds.): Semiconductors and semimetals, Vol. III, p. 17. New York: Academic Press 1967.

187. Yamada, E.: Proc. Int. Conf. Physics of Semiconductors, p. 274. Miasek, M. (Ed.).
     Warszawa: Polish Scientific Publishers 1972.
     Yamada, E., Kurosawa, T.: J. Phys. Soc. Japan **34**, 603 (1973).
188. Peterson, R. L.: Phys. Rev. B**5**, 3994 (1972); Phys. Rev. B**2**, 4135 (1970).
     Peterson, R. L., Magnusson, B., Weissglas, P.: Phys. Stat. Sol. (b) **46**, 729 (1971).
189. Pomortsev, R. V.: Sov. Phys. Solid State **9**, 1074 (1967).
190. Rode, D. L.: Phys. Rev. B**2**, 1012 (1970).
191. Kocevar, P.: Private communication.
192. See Ref. [10], p. 187.
193. Heinrich, H., Kriechbaum, M.: J. Phys. Chem. Solids **31**, 927 (1970).
194. See Ref. [10], p. 154.
195. Seeger, K., Bauer, G., Kuchar, F., Kuzmany, H.: Acta Phys. Austriaca **35**, 195 (1972).
196. Ruch, J. G.: Appl. Phys. Letters **20**, 246 (1972).
197. Kriechbaum, M., Heinrich, H., Wajda, J.: J. Phys. Chem. Solids **33**, 829 (1972).
     Asche, M., Sarbei, O. G., Savylow, Yu. G.: Phys. Stat. Sol. (b) **52**, 707 (1972).
     Asche, M., Sarbei, O. G.: Phys. Stat. Sol. **37**, 439 (1970).
198. Dienys, V., Pozhela, Yu. K.: Phys. Stat. Sol. **17**, 769 (1966).
199. Nathan, M. I.: Phys. Rev. **130**, 2201 (1963).
200. Schweitzer, D., Seeger, K.: Z. Phys. **183**, 207 (1965).
201. Baynham, A. C., Butcher, P. N., Fawcett, W., Loveluck, J. M.: Proc. Phys. Soc. (London)
     92, 783 (1967).
202. Hammar, C., Weissglas, P.: Phys. Stat. Sol. **24**, 531 (1967).
203. Hilsum, C.: Proc. Int. Conf. Physics of Semiconductors, p. 585. Miasek, M. (Ed.).
     Warszawa: Polish Scientific Publishers 1972.
204. Ruch, J. G., Fawcett, W.: J. Appl. Phys. **41**, 3843 (1970).
205. Fawcett, W., Paige, E. G. S.: J. Phys. C (Solid State Physics) **4**, 1801 (1971).
     Hammar, C.: Phys. Rev. B**4**, 417 (1971).
206. Hillbrand, H.: J. Phys. C (Solid State Physics) **5**, 3491 (1972).
     Löschner, H., Zimmerl, O., Frank, K., Hillbrand, H., König, W., Pötzl, H.: See Ref.
     [203], p. 630.
207. Hammar, C.: J. Phys. C (Solid State Phys.) **6**, 70 (1973).
208. Persky, G., Bartelink, D. J.: Phys. Rev. B**1**, 1614 (1970).
209. Maeda, H., Kurosawa, T.: See Ref. [203], p. 602.

Dr. Günther Bauer
I. Physikalisches Institut der Rheinisch-Westfälischen
Technischen Hochschule Aachen
D-5100 Aachen
Schinkelstr. 2
Federal Republic of Germany

Surface and Bulk Phonon-Polaritons
Observed by Attenuated Total Reflection

G. BORSTEL, H. J. FALGE, and A. OTTO

Contents

1. Introduction 107
2. Theory of Phonon-Polaritons in Infinite Crystals 108
3. Theory of Surface Phonon-Polaritons in Semiinfinite Crystals 111
   3.1. Basic Equations 111
   3.2. The Electromagnetic Field of the Surface Phonon-Polaritons 114
   3.3. TM-Polarized Surface Phonon-Polaritons in Orthorhombic Crystals 116
   3.4. TM-Polarized Surface Phonon-Polaritons in Uniaxial Crystals 118
4. Relation between TM-Reflectivity and Bulk Phonon-Polaritons 122
   4.1. Reflectivity Formulae 122
   4.2. Experimental Observation of TM-Reststrahlbands 127
5. Optical Excitation of Surface Waves 134
   5.1. Ordinary Surface Phonon-Polaritons 137
   5.2. Extraordinary Surface Phonon-Polaritons 141
References 145

Abstract. The basic equations of phonon-polaritons are reviewed and their relation to surface phonon-polaritons in semiinfinite crystals is given. Their observation by attenuated total reflection (ATR) and the resulting information on phonons are discussed.

1. Introduction

Polariton is the general term for coupled polarization excitation-photon modes. Consequently by phonon-polaritons the coupling of optical lattice vibrations as elementary excitations and photons is designated. Surface phonon-polaritons are interface modes and depend on the optical properties of at least two dielectric media, whereas phonon-polaritons propagate in an infinite ionic crystal (bulk polaritons). Since phonon-polaritons will be the subject of an extensive treatment in [1] here we mainly deal with surface phonon-polaritons. They are determined by the basic equations of phonon-polaritons and the electromagnetic boundary conditions. Due to the latter they differ in their physical properties for various shape and size of the crystal being investigated. We restrict ourselves to surface phonon-polaritons of semiinfinite crystals, which can be measured by a special ATR-arrangement. This method originally had been devised by Otto [2] for the observation of surface plasmon-polaritons. However, it turned out to be appropriate for the investigation of all dipole excitations in a plane surface and has been applied on surface phonon-polaritons during the last two years. Now the results are in a state where most of the basic questions are cleared up and may be used to determine and assign optical phonons. Hitherto this problem could hardly be solved for polyatomic crystals. New experimental methods are therefore desirable.

Moreover ATR allows the dispersion of phonon-polaritons to be measured. Since numerous optical phonons are only infrared (IR)-active, these modes could not in the past be investigated by Raman scattering. Now the corresponding spectra may be obtained by the attenuation of p-polarized (TM) radiation. In addition, they give a rather precise determination of IR-active optical phonons.

2. Theory of Phonon-Polaritons in Infinite Crystals

The theory of phonon-polaritons in infinite isotropic diatomic crystals (e.g. NaCl, ZnS) originally was developed by Huang [3] and was published later in a well known textbook on lattice dynamics by Born and Huang [4]. The extension of the theory to anisotropic crystals with an arbitrary number of atoms in the unit cell is essentially the work of Barker [5], Loudon [6, 7], and Merten [8–10]. In the meantime review papers on this subject have been published by Pick [11], Scott [12], Barker and Loudon [13], and Merten [14, 15]. For this reason we may confine ourselves to point out only the essential features of phonon-polaritons in infinite crystals. In the following we use the macroscopic phenomenological approach. For a microscopic approach based on quantum field theory the reader is referred to Pick [11].

The basic equations of long optical lattice waves in biaxial crystals[1] in the framework of the harmonic approximation are

$$\ddot{Q}' = B'^{11} Q' + B'^{12} E' , \qquad (2.1\,\text{a})$$

$$P' = (B'^{12})^+ Q' + B'^{22} E' , \qquad (2.1\,\text{b})$$

with the following notations: E': electric field, P': electric polarization, Q': vector of the quasi-normal coordinates with dimension r where r is the number of IR-active lattice vibrations, B'^{11}: $(r \times r)$-matrix, B'^{12}: $(r \times 3)$-matrix, B'^{22}: (3×3)-matrix, $(B'^{12})^+$: transposed matrix of B'^{12}.

The coefficients B'^{ik} are assumed to be constant, i.e. the influence of spatial dispersion is neglected. This assumption restricts the validity of

---

[1] It is assumed that the crystal axes are fixed by crystal symmetry, i.e. triclinic and monoclinic crystals must be excluded from the theory. In the following all vectors and tensor components which refer to these crystal axes are marked by a dash.

the theory to the region $k \approx 0$ of the Brillouin-zone ($k \lesssim 10^5$ cm^{-1}) which essentially is the range observable in common light scattering experiments.

Assuming $\exp(-i\omega t)$ time dependence and eliminating Q' from Eqs. (2.1 a) and (2.1 b) we obtain

$$P'_\alpha = \left\{ B'^{22}_{\alpha\alpha} + \sum_j (B'^{12}_{j\alpha})^2 \, (-B'^{11}_{(\alpha)jj} - \omega^2)^{-1} \right\} E'_\alpha . \tag{2.2}$$

α denotes the crystal axes and j the IR-active modes. Since $D' = E' + 4\pi P' = \varepsilon' E'$ (D': dielectric displacement, ε': dielectric tensor) the α-component of the dielectric tensor is

$$\varepsilon'_\alpha(\omega) = 1 + 4\pi B'^{22}_{\alpha\alpha} + \sum_j 4\pi (B'^{12}_{j\alpha})^2 \, (-B'^{11}_{(\alpha)jj} - \omega^2)^{-1} . \tag{2.3}$$

This expression for $\varepsilon'_\alpha(\omega)$ may be rearranged with respect to its zeros and poles. Denoting the zeros by $\omega^{LO}_{\alpha j}$ and the poles by $\omega^{TO}_{\alpha j}$ we obtain

$$\varepsilon'_\alpha(\omega) = \varepsilon'^{\infty}_\alpha \prod_j \frac{(\omega^{LO}_{\alpha j})^2 - \omega^2}{(\omega^{TO}_{\alpha j})^2 - \omega^2} , \tag{2.4}$$

where $\varepsilon'^{\infty}_\alpha$ denotes the high frequency dielectric constant in α-direction. Equation (2.4) has the form of an extended Kurosawa relation [16]. Obviously $\omega^{LO}_{\alpha j}$ and $\omega^{TO}_{\alpha j}$ are just the frequencies of the pure longitudinal and pure transverse long optical phonons in α-direction.

In the limiting case $\omega \to 0$ a generalized Lyddane-Sachs-Teller (LST)-relation [17, 18]

$$\varepsilon'_\alpha(0) = \varepsilon'^{\infty}_\alpha \prod_j (\omega^{LO}_{\alpha j})^2 / (\omega^{TO}_{\alpha j})^2 \tag{2.5}$$

is obtained which may be used to check the accuracy of experimental LO- and TO-data since the static dielectric constant $\varepsilon'_\alpha(0)$ may be determined by other independent experimental methods.

By means of Eq. (2.4) it is possible to determine the coefficients B'^{ik} of the basic Eq. (2.1)

$$B'^{11}_{(\alpha)jj} = -(\omega^{TO}_{\alpha j})^2 ,$$

$$(B'^{12}_{j\alpha})^2 = (\varepsilon'^{\infty}_\alpha / 4\pi) \frac{\prod\limits_i [(\omega^{LO}_{\alpha i})^2 - (\omega^{TO}_{\alpha j})^2]}{\prod\limits_{i \neq j} [(\omega^{TO}_{\alpha i})^2 - (\omega^{TO}_{\alpha j})^2]} , \tag{2.6}$$

$$B'^{22}_{\alpha\alpha} = (\varepsilon'^{\infty}_\alpha - 1)/4\pi .$$

Thus all these coefficients may be determined by measurements of the frequencies of transverse and longitudinal[2] long optical phonons and

---

[2] Instead of $\omega^{LO}_{\alpha j}$ the oscillator strengths $\varrho_{\alpha j}$ may be determined since the relation $(B'^{12}_{j\alpha})^2 = \varrho_{\alpha j} (\omega^{TO}_{\alpha j})^2$ holds.

the high frequency dielectric constants for the principal directions of
the crystal.

To describe phonon-polaritons the basic Eq. (2.1) of long optical
lattice modes must be combined with Maxwell's equations

$$\text{div } \boldsymbol{D}' = 0, \tag{2.7a}$$

$$\text{div } \boldsymbol{H}' = 0, \tag{2.7b}$$

$$\text{curl } \boldsymbol{E}' = -(1/c)\,\dot{\boldsymbol{H}}', \tag{2.7c}$$

$$\text{curl } \boldsymbol{H}' = (1/c)\,\dot{\boldsymbol{D}}'. \tag{2.7d}$$

Herein $\boldsymbol{H}'$ denotes the magnetic field, c the velocity of light in vacuum.

Assuming $\exp(i\boldsymbol{k}' \cdot \boldsymbol{x}')$ spatial dependence ($\boldsymbol{k}'$: wave vector), Maxwell's
equations (2.7) may be combined to

$$\boldsymbol{D}' = (c^2/\omega^2)\left((\boldsymbol{k}' \cdot \boldsymbol{k}')\,\boldsymbol{E}' - (\boldsymbol{k}' \cdot \boldsymbol{E}')\,\boldsymbol{k}'\right) \tag{2.8}$$

which is a well known relation of crystal optics.

Since $\boldsymbol{D}' = \varepsilon'\boldsymbol{E}'$, where ε' is given by (2.3) or (2.4), we obtain from
(2.8) explicitly

$$\begin{bmatrix} k_y'^2 + k_z'^2 - (\omega^2/c^2)\varepsilon_x' & -k_x'k_y' & -k_x'k_z' \\ -k_x'k_y' & k_x'^2 + k_z'^2 - (\omega^2/c^2)\varepsilon_y' & -k_y'k_z' \\ -k_x'k_z' & -k_y'k_z' & k_x'^2 + k_y'^2 - (\omega^2/c^2)\varepsilon_z' \end{bmatrix} \begin{bmatrix} E_x' \\ E_y' \\ E_z' \end{bmatrix} = 0. \tag{2.9}$$

The eigenfrequencies $\omega(\boldsymbol{k}')$ of polaritons are determined by the condition
that (2.9) has a nontrivial solution only if the corresponding determinant
vanishes. As can be shown [9] this condition leads to an implicit dispersion relation in form of a generalized Fresnel-equation

$$F(k_x', k_y', k_z', \omega) \equiv \sum_\alpha \frac{\varepsilon_\alpha'(\omega)\,k_\alpha'^2}{n^2 - \varepsilon_\alpha'(\omega)} = 0, \tag{2.10}$$

where $n^2 \equiv (c^2/\omega^2)\,(\boldsymbol{k}' \cdot \boldsymbol{k}')$ denotes the refractive index. Evaluation of
(2.10) today is a standard procedure for calculating polariton dispersion
curves of arbitrary biaxial crystals. It should be emphasized that in
anisotropic crystals the frequency ω of polaritons is dependent both on
the modulus and the direction of the wave vector $\boldsymbol{k}'$.

Pure transverse phonon-polaritons ($\boldsymbol{k}' \perp \boldsymbol{E}'$) occur if $\boldsymbol{k}'$ is along a
principal axis α of the dielectric tensor of the crystal and in addition
$n^2 - \varepsilon_\beta' = 0$ ($\beta \neq \alpha$) holds. From (2.9) it follows that the only nonzero
component of $\boldsymbol{E}'$ is E_β'. Pure longitudinal modes ($\boldsymbol{k}' \parallel \boldsymbol{E}'$) may only occur
in the same geometry but now we have $\varepsilon_\alpha'(\omega) = 0$. From (2.9) and (2.7)
we obtain $\boldsymbol{D}' = \boldsymbol{H}' = 0$, $\text{curl } \boldsymbol{E}' = 0$. Thus pure longitudinal phonon-
polaritons are not coupled to the electromagnetic field and show no
dispersion according to the modulus of $\boldsymbol{k}'$.

The directional dispersion of long optical phonons may be evaluated from the limiting case $k' \to \infty$ $(n \to \infty)$ of Eq. (2.10). It follows

$$F(s'_x, s'_y, s'_z, \omega) \equiv \sum_\alpha \varepsilon'_\alpha(\omega)\, s'^2_\alpha = 0 \tag{2.11}$$

where $s' \equiv k'/|k'|$ denotes the wave normal vector. Comparing with Maxwell's equations (2.7), this is the electrostatic approximation (curl $E' = 0$) and equivalent to the limiting case $k' \to 0$ in the usual electrostatic theory of lattice vibrations (Born-v. Kármán theory).

For uniaxial crystals Fresnel's equation (2.10) splits into two dispersion relations for the ordinary and extraordinary polaritons, respectively. Denoting vector and tensor components parallel to the optic axis by the index $\parallel$, those perpendicular to it by the index $\perp$, we find

$$n^2 - \varepsilon_\perp = 0 \quad \text{(ordinary polaritons)}, \tag{2.12a}$$

$$\varepsilon_\perp(n^2 - \varepsilon_\parallel)\, k^2_\perp + \varepsilon_\parallel(n^2 - \varepsilon_\perp)\, k^2_\parallel = 0 \quad \text{(extraordinary polaritons)}. \tag{2.12b}$$

Ordinary polaritons are pure transverse waves and show no directional dispersion, whereas extraordinary polaritons in general show mixed polarization and are dependent on both the direction and the modulus of the wavevector $k' = (k_\perp, k_\parallel)$.

In order to determine the eigenvectors E', P', Q' of phonon-polaritons the homogeneous systems of Eqs. (2.9) and (2.1) may be solved numerically. This approach has been used by Borstel and Merten [19, 20] and Unger and Schaack [21, 22]. In a recent paper Borstel and Merten [23] have shown that the eigenvectors also may be evaluated analytically. Their result for the eigenvector E' of (2.9) is

$$E'_\alpha = \frac{k'_\alpha}{k'_\beta} \frac{n^2 - \varepsilon'_\beta}{n^2 - \varepsilon'_\alpha} E'_\beta, \tag{2.13}$$

where E'_β may be considered as a normalization constant. The remaining eigenvectors P' and Q' then may be determined from (2.1). The knowledge of the eigenvectors is necessary to evaluate Raman intensities of polaritons [15, 24]. For further details concerning the calculation of such intensities the reader is referred to [22] and [23].

3. Theory of Surface Phonon-Polaritons in Semiinfinite Crystals

3.1. Basic Equations

After the development of the macroscopic theory of phonon-polaritons in infinite isotropic crystals the question soon arose as to whether its conclusions, e.g. the well known LST-relation, would also hold in semiinfinite or finite crystals. To our knowledge Rosenstock [25] was the first who pointed out the problem of the appropriate choice of boundary conditions in the lattice dynamics of finite crystals. He showed that the application of the common cyclic boundary conditions of lattice dynamics, which involves the assumption that every ion is equivalent to every other ion of the same species, for finite crystals with long range Coulomb interactions necessarily yields physically incorrect results. In fact all attempts to assume the ionic crystal to be semiinfinite or finite and at the same time to retain cyclic boundary conditions led to contradictory predictions [26, 27].

The problem was solved in 1965 when Fuchs and Kliewer [28–31] and later Englman and Ruppin [32–34], based on considerations of Barron [35], showed that the correct boundary conditions were those dictated by Maxwell's equations. It was found that for crystals with typical dimensions of at least several micrometers, the variation of interionic distances and short range force constants for atoms near the crystal surface may be neglected and that the macroscopic theory of long waves due to Huang, combined with classical electromagnetic theory, can give a full explanation of the optical properties of semiinfinite and finite crystals. For a review of these theories which essentially cover isotropic crystals the reader is referred to Ruppin and Englman [36].

The appropriate formalism for short waves or microcrystals with dimensions in the Ångström region has been developed by Feuchtwang [37, 38], and later by Lucas [39], Tong and Maradudin [40], Martin [41], and Trullinger and Cunningham [42] on the basis of different microscopic models. Since all these theories do not take into account the retardation of the Coulomb interaction, the validity of these theories is restricted to the region $k \gtrsim 10^5 \, \mathrm{cm}^{-1}$ of the Brillouin-zone and thus can only give information about light scattering experiments in the limiting case of long optical phonons.

The development of the ATR-method by Otto [2] and its application to the measurement of surface phonon-polaritons in cubic and uniaxial crystals has stimulated a great deal of theoretical work to extend the macroscopic theory of phonon-polaritons to semiinfinite and finite crystals of uniaxial and biaxial symmetry. This extension has been given in recent years by Lyubimow and Sannikov [43], Hartstein et al. [44–46], and Borstel [47–49]. It is the aim of this chapter to review the essential conclusions of these authors with respect to surface phonon-polaritons in semiinfinite uniaxial and biaxial crystals.

To investigate phonon-polaritons propagating along a plane interface between two dielectrics the geometry shown in Fig. 1 is chosen. An orthorhombic crystal characterized by a frequency dependent dielectric

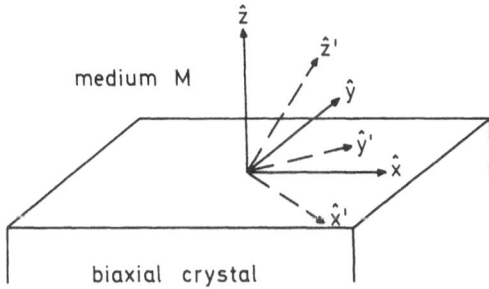

Fig. 1. Geometry for surface phonon-polaritons in biaxial crystals

tensor $\varepsilon(\omega)$ fills the lower half space $z < 0$, the upper half space $z > 0$ consists of an isotropic medium M with dielectric constant $\varepsilon_M \geqq 1$.

The unit vectors $\hat{x}$, $\hat{y}$, $\hat{z}$ build up an orthonormal basis B of the interface at $z = 0$. $\hat{x}'$, $\hat{y}'$, $\hat{z}'$ denote the principal axes of the dielectric tensor $\varepsilon(\omega)$ of the crystal. They form a second orthonormal system B' and may be oriented in arbitrary directions with respect to B.

The connection of B and B' is described by an orthogonal transformation $\mathbf{T}$:

$$\mathbf{T}: B \to B', \qquad \mathbf{T}^+: B' \to B.$$

The operator $\mathbf{T}$ is completely determined by the crystal cut and its components may therefore be regarded as known quantities.

Since we are dealing with the long wave limit $k \approx 0$ ($k \lesssim 10^5$ cm^{-1}) the surface modes of the semiinfinite crystal describe the motion of ions which are essentially in the bulk of the crystal. The equations of motion of the lattice therefore are those of the infinite crystal, i.e. Huang's equations (2.1) will hold in this case too. Obviously all conclusions of Chapter 2 which follow only from these two equations (i.e. without inclusion of Maxwell's equations) remain valid for semiinfinite crystals. In particular this means that the dielectric functions $\varepsilon'_\alpha(\omega)$ may be represented in the form (2.4) and that the generalized LST-relation (2.5) also holds in semiinfinite samples.

In order to describe phonon-polaritons we have to combine the system (2.1) with Maxwell's equations (2.7). This leads to the dispersion of the bulk phonon-polaritons as given by Eq. (2.10).

For surface phonon-polaritons, one chooses solutions of Eqs. (2.1) and (2.7), which are periodic along the interface between the orthorhombic crystal and the isotropic medium M. These fields have to obey the electromagnetic boundary conditions at $z = 0$. The fields must disappear far from the interface for large $|z|$.

The periodicity is given for all electromagnetic field components – $(E, D, H$ in the crystal and E_M, D_M, H_M in M) – by the phasefactor

$$\exp\{i(k_x x + k_y y - \omega t)\} \tag{3.1}$$

where the two dimensional vector (k_x, k_y) is the wave-vector of the surface phonon-polariton. The boundary conditions of the electromagnetic fields at $z = 0$ are

$$\hat{z} \times (E - E_M) = 0 \tag{3.2a}$$
$$\hat{z} \times (H - H_M) = 0$$

and

$$\hat{z} \cdot (D - D_M) = 0 \tag{3.2b}$$
$$\hat{z} \cdot (H - H_M) = 0 .$$

We note at this point, that in the macroscopic theory outlined here the influence of the crystal surface is taken into account only by these boundary conditions, since in the long wavelength limit all microscopic surface effects are negligible. It is easy to prove that the boundary conditions (3.2b) follow from the conditions (3.2a).

3.2. The Electromagnetic Field of the Surface Phonon-Polaritons

The variation of the field of the surface polariton of wavevector (k_x, k_y) is easily calculated in medium M.

It is given by $\exp(ik_{Mz}z)$ with k_{Mz} obtained from Eq. (3.3)

$$k_{Mz}^2 + k_x^2 + k_y^2 = (\omega/c)^2 \varepsilon_M . \tag{3.3}$$

When there is no damping in the medium M and the orthorhombic crystal – e.g. when the harmonic approximation holds – the electromagnetic energy of the surface phonon-polariton is conserved in time. Thus the time averaged energy flow across the surface must be zero. From the expression

$$S = (1/2)(c/4\pi) E \times H^* \tag{3.4}$$

for the complex Poynting vector one derives for the z-component [which is continuous across the boundary, as follows from (3.2a)]

$$\operatorname{Re} S_z = \operatorname{Re} S_{Mz} = (c^2/8\pi\omega)|E_M|^2 \operatorname{Re} k_{Mz} .$$

In order that $\mathrm{Re}\,S_z$ vanishes it is sufficient that

$$\mathrm{Re}\,k_{Mz} = 0. \tag{3.5}$$

Together with (3.3) this leads to the condition of the existence of a surface polariton

$$k_x^2 + k_y^2 > (\omega^2/c^2)\,\varepsilon_M. \tag{3.6}$$

Within the orthorhombic crystal, the electromagnetic field

$$F(x, y, z) = F(z)\exp\{i(k_x x + k_y y - \omega t)\} \tag{3.7}$$

of the crystal in general cannot be described by the simple ansatz with a wavevector component k_z. Usually the solution of (3.7), (2.1), and (2.7) will yield fields of the form

$$F(z) = F^{(1)}\exp(ik_z^{(1)}z) + F^{(2)}\exp(ik_z^{(2)}z). \tag{3.8}$$

The fields of this form are characterized by two independent field amplitudes, the ratio of which has to be evaluated by the boundary conditions (3.2a). The surface phonon-polariton field in the medium M may be separated unambiguously into a transverse magnetic (TM) and transverse electric (TE) component. The two field variables in M and the two in the crystal have to fulfil the 4 boundary conditions (3.2a). For a nontrivial solution, the determinant of the homogeneous system of equations for the boundary conditions must be zero. This yields the dispersion relation of the surface phonon-polariton. This problem has been solved by Lyubimov and Sannikov [43] for an uniaxial crystal. For the general case, there exists no analytical solution for the dispersion, it is rather implicitly contained in a lengthy, cumbersome expression.

In view of the experimental application of the theory, this general case is of no great importance, for two reasons:

a) For a quantitative experimental analysis of a surface phonon-polariton resonance in an attenuated total reflection spectrum, ellipticity of the reflected light has to be measured. This requires a polarizer also in the reflected beam – there may be problems of intensity and of calibration of the intensities of different polarization. Only for TM-polarized surface phonon-polaritons the ellipticity is zero and hence no polarizer is needed in the reflected beam.

b) The full information on the phonon structure of the orthorhombic crystal may be already obtained from pure TM-polarized surface phonon-polaritons.

Therefore, the next section will treat the conditions for the existence of pure TM-polarized surface phonon-polaritons.

3.3. TM-Polarized Surface Phonon-Polaritons in Orthorhombic Crystals

Both fields $F^{(1)}$ and $F^{(2)}$ in Eq. (3.8) obey the wave equation. For the electric field vector E, it is given by

$$k^2 E - (k \cdot E) k = (\omega^2/c^2) \varepsilon E \tag{3.9}$$

where $k = (k_x, k_y, k_z)$. The first two components of k constitute the two dimensional wavevector of the surface phonon-polariton. ε is the dielectric tensor in the coordinate system $\hat{x}, \hat{y}, \hat{z}$, which is related to the tensor

$$\varepsilon' = \begin{bmatrix} \varepsilon'_x & 0 & 0 \\ 0 & \varepsilon'_y & 0 \\ 0 & 0 & \varepsilon'_z \end{bmatrix} \tag{3.10}$$

by

$$\varepsilon = T^+ \varepsilon' T . \tag{3.11}$$

Without loss of generality, we may assume $k_y = 0$. We may include the additional rotation of the coordinate system, which brings the x-axis parallel to the two dimensional wavevector of the surface phonon-polariton, into the rotation T.

A TM surface phonon-polariton with $k_y = 0$ is characterized by $E_y = 0$, $H_x = H_z = 0$, as follows immediately from the condition $H_z = 0$.

A solution with $E_y = 0$ is only compatible with Eq. (3.9), when ε has the form

$$\varepsilon = \begin{bmatrix} \varepsilon_x & 0 & \varepsilon_{xz} \\ 0 & \varepsilon_y & 0 \\ \varepsilon_{xz} & 0 & \varepsilon_z \end{bmatrix} . \tag{3.12}$$

From (3.11) and (3.12) follows, that one of the dielectric tensor axes $\hat{x}', \hat{y}', \hat{z}'$ must coincide with $\hat{y}$. That means, for the existence of TM surface phonon-polaritons one dielectric tensor axis must be parallel to the surface and perpendicular to the direction of the surface polariton wavevector.

From $E_y = 0$ and $\mathrm{Re} S_z = 0$ one derives immediately, that for TM surface phonon-polaritons the time averaged Poynting vector is parallel or antiparallel to the surface phonon-polariton wavevector.

With ε given by (3.12), the calculation of k_z from (3.9) yields a quadratic equation with

$$k_z = - k_x(\varepsilon_{xz}/\varepsilon_z) \pm (1/\varepsilon_z) (k_x^2 - \omega^2 \varepsilon_z/c^2)^{1/2} (- \varepsilon'_x \varepsilon'_z)^{1/2} . \tag{3.13}$$

For the general case, (3.9) leads to a 4th order equation in k_z. For TM surface phonon-polaritons, one choses the k_z value from (3.13), for which the fields vanish for large $|z|$. In this case, one has only one field F within the crystal rather than two as given in Eq. (3.8).

For the TM case, the four boundary conditions (3.2a) reduce to two boundary conditions, namely that E_x and H_y are continuous across the interface. With only one field amplitude in M and one in the crystal this allows for a consistent solution of the TM surface phonon-polariton dispersion.

Regular TM-polarized surface polaritons are characterized by a purely exponential decay within the solid, whereas non regular ones decay like the product of an exponential and a sinusoidal function.

This decay is given by $\exp(i k_z z)$, with k_z from Eq. (3.13). For a regular surface polariton k_z must be negative imaginary. As all quantities in (3.13) are real, this is only possible if $\varepsilon_{xz} = 0$.

Therefore, regular TM-polarized surface phonon-polaritons only exist on surfaces perpendicular to one dielectric axis, and a second dielectric axis parallel to the wavevector of the surface polariton.

For the regular TM-polarized surface phonon-polaritons, an explicit dispersion relation is obtained. It is given by [44]

$$k_x^2 = (\omega^2/c^2)\varepsilon_M \varepsilon_z'(\varepsilon_x' - \varepsilon_M)/(\varepsilon_x'\varepsilon_z' - \varepsilon_M^2) \tag{3.14}$$

with the condition $\varepsilon_x' < 0$.

It should be emphasized that the dispersion of TM-polarized surface phonon-polaritons in orthorhombic crystals is described already by two components of the dielectric tensor ε'. As will be seen in the next section for that reason there is a strong analogy between these modes and the extraordinary surface phonon-polaritons of uniaxial crystals.

For isotropic crystals ($\varepsilon_x' = \varepsilon_y' = \varepsilon_z' \equiv \varepsilon$) we obtain from (3.14) the well known dispersion relation

$$(k_x^2 c^2)/(\omega^2) = (\varepsilon_M \varepsilon)/(\varepsilon + \varepsilon_M) \tag{3.15}$$

of surface phonon-polaritons in semiinfinite cubic crystals. The corresponding attenuation constants k_z and k_{Mz} are

$$k_z = (\omega/c)\left[\varepsilon^2/(\varepsilon + \varepsilon_M)\right]^{1/2}, \qquad k_{Mz} = (\omega/c)\left[\varepsilon_M^2/(\varepsilon + \varepsilon_M)\right]^{1/2}. \tag{3.16}$$

These equations hold for every direction of propagation of the mode parallel to the surface and for every crystal cut. Obviously the surface waves which may be shown to be pure transverse modes exist only in frequency regions where $-\infty < \varepsilon(\omega) < -\varepsilon_M$. For a diatomic cubic crystal the dispersion curve is sketched in Fig. 2. Starting at the point $P(\omega = \omega^{TO})$ the branch approaches in the limiting case $k_x \to \infty$ the frequency ω^s given by

$$\varepsilon(\omega^s) = -\varepsilon_M. \tag{3.17}$$

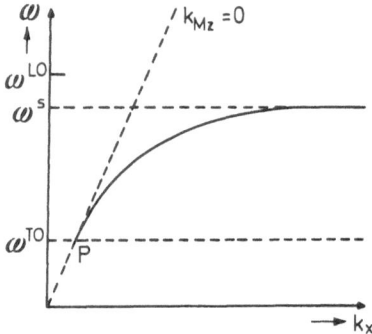

Fig. 2. Dispersion of surface phonon-polaritons in a semiinfinite cubic crystal with two atoms

As may be seen from (3.16), in the point P the surface phonon-polariton is merely a photon in M, travelling parallel to the surface in direction of k_x.

3.4. TM-*Polarized Surface Phonon-Polaritons in Uniaxial Crystals*

The results of the previous section may be easily specialized to uniaxial crystals. For TM surface phonon-polaritons, propagating in x-direction, one dielectric axis has to point into the y-direction (see Fig. 1). This is the case for uniaxial crystals, when the optical axis is in y-direction or in the $x - z$ plane (Fig. 3).

For the first case, one obtains regular surface phonon-polaritons with the explicit dispersion relation [43, 46]

$$k_x^2 = (\omega^2/c^2)\,(\varepsilon_\perp \cdot \varepsilon_M)/(\varepsilon_\perp + \varepsilon_M)\,. \qquad (3.18)$$

This has the same form as the dispersion relation (3.16) of surface phonon-polaritons in isotropic crystals. In analogy to ordinary bulk polaritons,

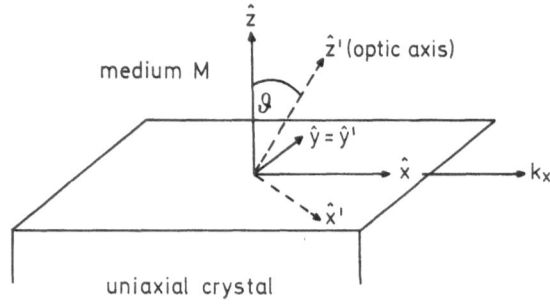

Fig. 3. Geometry for surface phonon-polaritons in uniaxial crystals

the TM surface phonon-polaritons of this dispersion have been labelled "ordinary". This dispersion behaviour is of the type sketched in Fig. 2. Ordinary surface phonon-polaritons are apart from the surface polarization charge transverse excitations, because the divergence of the electric field is zero within M and within the crystal.

For the case depicted in Fig. 3, we have

$$\mathbf{T} = \begin{bmatrix} \cos\vartheta & 0 & -\sin\vartheta \\ 0 & 1 & 0 \\ \sin\vartheta & 0 & \cos\vartheta \end{bmatrix}. \tag{3.19}$$

With

$$\boldsymbol{\varepsilon}' = \begin{bmatrix} \varepsilon_\perp & 0 & 0 \\ 0 & \varepsilon_\perp & 0 \\ 0 & 0 & \varepsilon_{||} \end{bmatrix} \tag{3.20}$$

we obtain from Eq. (3.11) for the components of (3.12)

$$\varepsilon_x = \varepsilon_\perp \cos^2\vartheta + \varepsilon_{||} \sin^2\vartheta; \quad \varepsilon_z = \varepsilon_\perp \sin^2\vartheta + \varepsilon_{||} \cos^2\vartheta;$$

$$\varepsilon_y = \varepsilon_\perp; \quad \varepsilon_{xz} = (\varepsilon_{||} - \varepsilon_\perp) \sin\vartheta \cos\vartheta. \tag{3.21}$$

Explicit dispersion relations of the TM surface phonon-polariton may be obtained for $\vartheta = 0$ and $\vartheta = \pi/2$.

a) For $\vartheta = 0$ the physical properties of the sample are invariant for rotations around the surface normal $\hat{z}$. The dispersion is given by [43, 46]

$$k_x^2 + k_y^2 = (\omega^2/c^2)\,\varepsilon_M \varepsilon_{||}(\varepsilon_\perp - \varepsilon_M)/(\varepsilon_{||}\varepsilon_\perp - \varepsilon_M^2). \tag{3.22}$$

b) For $\vartheta = \pi/2$, one obtains [43, 46]

$$k_x^2 = (\omega^2/c^2)\,\varepsilon_M \varepsilon_\perp(\varepsilon_{||} - \varepsilon_M)/(\varepsilon_{||}\varepsilon_\perp - \varepsilon_M^2). \tag{3.23}$$

According to the conclusions of Section 3.3, these TM surface phonon-polaritons are regular. They have a longitudinal component, as the divergence of the electric field in the crystal does not disappear. In analogy to the extraordinary bulk polaritons, they have been labelled extraordinary.

Non regular TM surface phonon-polaritons are possible for $0 < \vartheta < \pi/2$ and $k_y = 0$. The dispersion behaviour is described by the Eq. (3.13) [the tensor components have to be taken from (3.21)], (3.24) and (3.25)

$$(k_x^2 + k_z^2 - \omega^2 c^{-2}\varepsilon_{||})\,(\cos\vartheta \cdot k_x - \sin\vartheta \cdot k_z)\,(\varepsilon_M \cos\vartheta \cdot k_x - \varepsilon_\perp \sin\vartheta \cdot k_{Mz}) \tag{3.24a}$$

$$+ (k_x^2 + k_z^2 - \omega^2 c^{-2}\varepsilon_\perp)\,(\sin\vartheta \cdot k_x + \cos\vartheta \cdot k_z)\,(\varepsilon_M \sin\vartheta \cdot k_x + \varepsilon_{||} \cos\vartheta \cdot k_{Mz}) = 0,$$

$$k_x^2 + k_{Mz}^2 = (\omega^2/c^2)\,\varepsilon_M, \tag{3.25}$$

which may not be combined to an explicit dispersion relation.

An equivalent form of Eq. (3.24 a) is

$$(k_x^2 + k_z^2)(\varepsilon_M k_x^2 + \varepsilon_z k_z k_{Mz} + \varepsilon_{xz} k_x k_{Mz})$$
$$- \omega^2 c^{-2}(\varepsilon_M \varepsilon_z k_x^2 + \varepsilon_{\parallel} \varepsilon_{\perp} k_z k_{Mz} - \varepsilon_{xz} \varepsilon_M k_x k_z) = 0 \,. \tag{3.24b}$$

Since (3.24) is not a pure quadratic in k_x the question arises as to whether there is any difference in the dispersion behaviour when k_x is changed to $-k_x$. Splitting (3.13) and (3.24) into their real and imaginary parts it is found that $\omega(k_x) = \omega(-k_x)$, i.e. an opposite direction of propagation of the surface mode does not affect the dispersion curve.

Although the Eqs. (3.13), (3.24), and (3.25) in general must be evaluated by means of a computer we may draw some essential conclusions concerning the dispersion of TM surface phonon-polaritons in a pure analytic way. Since surface phonon-polaritons are characterized by a nonvanishing imaginary part $\mathrm{Im}\, k_z$ of the normal component of k the discriminant of (3.13) must be negative. From (3.13) and (3.21) it is found

$$(\varepsilon_{\perp} \varepsilon_{\parallel}/\varepsilon_z^2)(\omega^2 \varepsilon_z/c^2 - k_x^2) < 0 \,. \tag{3.26}$$

The limits of the allowed frequency region thus are given by

$$\varepsilon_{\perp} \varepsilon_{\parallel}(\omega^2 c^{-2} \varepsilon_z - k_x^2) = 0 \,, \tag{3.27}$$

which corresponds to

$$\mathrm{Im}\, k_z = 0 \,. \tag{3.28}$$

On the other hand the region of allowed values of k_x has a lower limit given by

$$k_{Mz} = 0 \,, \tag{3.29}$$

which according to (3.25) is equivalent to

$$(c^2/\omega^2) k_x^2 - \varepsilon_M = 0 \,. \tag{3.30}$$

As the lefthand side of (3.26) changes sign at LO-phonons the dispersion branch of surface phonon-polaritons may not cross the corresponding lines in the (ω, k_x) diagram. This fact leads to the distinction of real and virtual excitation surface phonon-polaritons as shown for weak anisotropy in Fig. 4.

In Fig. 4a the curve $C(\vartheta)$ given by

$$k_x^2 = \omega^2 \varepsilon_z/c^2 \tag{3.31}$$

which represents a lower limit for the dispersion branch does not cross the line $\omega = \omega^{LO}$. Consequently the dispersion branch may extend to $k_x \to \infty$, i.e. in the electrostatic limiting case a long optical surface phonon

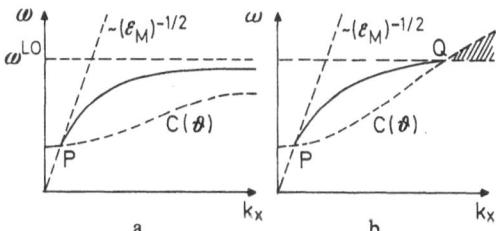

Fig. 4a and b. Dispersion of real (a) and virtual (b) excitation surface phonon-polaritons in semiinfinite uniaxial crystals [48]

will exist. Such surface modes are termed real excitation surface phonon-polaritons.

In Fig. 4b a crossing point Q does exist. It may be easily verified that Q is a solution of (3.24) and therefore a point of the dispersion curve.

Beyond Q (hatched area) surface phonon-polaritons may no longer exist since above $\omega = \omega^{LO}$ only bulk modes can occur. Consequently the dispersion branch will terminate in Q. Such surface excitations which do exist only for finite values of k_x are termed virtual excitation modes.

At the stop point Q the electromagnetic field of the surface phonon-polaritons within the crystal decays into a pure longitudinal phonon. and a pure transverse bulk polariton (polarized in the $x - z$ plane). Both phonon and polariton have the same k vector, which is either parallel or perpendicular to the optical axis.

The existence of virtual surface excitations is a characteristic property of anisotropic crystal samples. For isotropic crystal ($\varepsilon_\perp = \varepsilon_{||} = \varepsilon_z$) the antefactor $\varepsilon_\perp \varepsilon_{||}/\varepsilon_z^2$ of the lefthand side of (3.26) becomes equal to one and therefore cannot result in a change of sign of the discriminant.

The number of real excitation extraordinary surface phonon-polaritons may be obtained by an evaluation of the limiting case $k_x \to \infty$ for finite ω. From (3.24) it is found for the corresponding limiting frequencies ω_j^s

$$\varepsilon_\perp(\omega_j^s) \cdot \varepsilon_{||}(\omega_j^s) = \varepsilon_M^2 . \tag{3.32}$$

This equation shows that the limiting frequencies are independent of the crystal cut ϑ (Fig. 3). Since (3.32) requires $\varepsilon_\perp \varepsilon_{||} > 0$ and solutions for which $\varepsilon_\perp > 0$, $\varepsilon_{||} > 0$ must be excluded the total number of real surface phonon-polaritons is given by the number of solutions of (3.32) which fulfil $\varepsilon_\perp < 0$, $\varepsilon_{||} < 0$. For the existence of real extraordinary modes therefore an overlap of the negative regions of $\varepsilon_\perp$ and $\varepsilon_{||}$ is necessary. Thus in crystals for which anisotropy in the interatomic forces predominates over long range Coulomb forces real excitation TM surface phonon-polaritons will not exist. But if an overlap occurs then it is easy to see that the product $\varepsilon_\perp \varepsilon_\parallel$ in such a region may take every value between zero and infinity. The number of solutions of (3.32) therefore is independent of the dielectric constant ε_M of the outer medium, i.e. the total number of real excitation TM-polarized extraordinary surface phonon-polaritons is an invariant property of the uniaxial crystal.

The stop point of virtual excitation surface phonon-polaritons is given by the crosspoint of curve $C(\vartheta)$ [Eq. (3.31)] and a longitudinal optical phonon, and therefore does not depend on ε_M. In some experimental cases, ε_M may be chosen so large, that the k vector at this crosspoint does not fulfil (3.6). In these cases, the virtual excitation surface phonon-polariton is quenched.

4. Relation between TM-Reflectivity and Bulk Phonon-Polaritons

4.1. Reflectivity Formulae

The components of the dielectric tensor are the only properties of the material which are contained in the dispersion relations of ordinary and extraordinary phonon-polaritons. This is also the case in the well known formulae of reflection [52–62]. It can be shown that regions of high reflectivity in the (ω, k) diagram are surrounded by curves and lines which give the dispersion of phonon-polaritons and the frequencies of IR-active longitudinal and transverse optical phonons. Since numerous phonons are only IR-active or have too low an intensity to be observable by Raman scattering the corresponding phonon-polaritons cannot be investigated otherwise[3]. Direct IR measurements of phonons require thin films [64]. For anisotropic crystals in particular they hardly can be prepared. The evaluation of reflection spectra, which take advantage of attenuated total reflection (ATR), is therefore a comparatively convenient method for determining the frequencies of phonons. In the following we shall give the principles of the experiment and explain the interpretation of the spectra.

The upper half space in Fig. 5 consists of an isotropic medium M with refractive index n_M. An orthorhombic crystal fills the lower half space. The principal axes of its dielectric tensor are parallel to $\hat{x}, \hat{y}, \hat{z}$.

$$\varepsilon = \begin{bmatrix} \varepsilon_x & 0 & \\ 0 & \varepsilon_y & 0 \\ 0 & 0 & \varepsilon_z \end{bmatrix}.$$

---

[3] Only the uppermost phonon-polariton branch has been obtained by measuring thermal emission spectra [63].

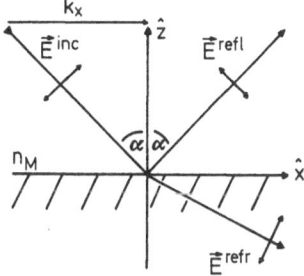

Fig. 5. Geometry for phonon-polaritons in biaxial crystals

At $z=0$ an incident electromagnetic wave (TM-polarized) is partly reflected and partly refracted. For the reflectivity, it follows from the boundary conditions that

$$R_{TM}(\alpha, \omega) = \left| \frac{\varepsilon_x^{1/2} \varepsilon_z^{1/2}(\omega/c)\cos\alpha - n_M[\varepsilon_z(\omega^2/c^2) - k_x^2]^{1/2}}{\varepsilon_x^{1/2} \varepsilon_z^{1/2}(\omega/c)\cos\alpha + n_M[\varepsilon_z(\omega^2/c^2) - k_x^2]^{1/2}} \right|^2 . \tag{4.1}$$

k_x is given by

$$k_x = (\omega/c)\, n_M(\omega)\sin\alpha \tag{4.2}$$

and may be varied by changing the angle of incidence α or by the selection of n_M.

The Eq. (4.1) may be transformed in the following way. The dispersion relation of the transverse bulk polariton with $\mathbf{k} \parallel \hat{x}$ and lattice displacements parallel to the direction $\hat{z}$ is

$$k_{BP}^2 = (\omega^2/c^2)\varepsilon_z \quad \text{(BP: Bulk polariton)}. \tag{2.10'}$$

The corresponding dispersion relation for extraordinary phonon-polaritons in uniaxial crystals (optical axis perpendicular to the surface) is

$$k_{BP}^2 = (\omega^2/c^2)\varepsilon_{\parallel}. \tag{2.12b'}$$

For ordinary phonon-polaritons in uniaxial crystals (optical axis parallel to $\hat{x}$ and $k_x \parallel \hat{x}$) we have

$$k_{BP}^2 = (\omega^2/c^2)\varepsilon_{\perp}. \tag{2.12a'}$$

From (2.10') Eq. (4.1) may be written as

$$R_{TM}(\alpha, \omega) = \left| \frac{\varepsilon_x^{1/2} k_{BP}\cos\alpha - n_M[k_{BP}^2 - k_x^2]^{1/2}}{\varepsilon_x^{1/2} k_{BP}\cos\alpha + n_M[k_{BP}^2 - k_x^2]^{1/2}} \right|^2 . \tag{4.3}$$

Neglecting damping, total reflection appears if one of the two terms which constitute the numerator and the denumerator in (4.3) – but not both – becomes imaginary. This condition requires for $\varepsilon_z > 0$ that either ε_x or $k_{BP}^2 - k_x^2$ is negative and for $\varepsilon_z < 0$ that ε_x is also negative. In all other cases there is decreased reflection. Then the z-component of the refracted ray

$$k_z^{refr} = \varepsilon_z^{-1} \left[(\varepsilon_z \omega^2 c^{-2} - k_x^2) \varepsilon_x \varepsilon_z \right]^{1/2} = \varepsilon_z^{-1} \left[(k_{BP}^2 - k_x^2) \varepsilon_x \varepsilon_z \right]^{1/2} \qquad (4.4)$$

is real and couples with the phonons propagating in z-direction. (4.4) is derived from Eq. (3.13). As can be seen for uniaxial crystals in Fig. 6 the ranges of total reflection (hatched areas) are surrounded by horizontal lines at the frequencies of phonons and the dispersion curves given by (2.12a') or (2.12b'). If damping is included the edges of calculated TM-reststrahlbands become smooth and the boundaries of the hatched areas can be found rather precisely by the turning points of the bands. The spectra can be recorded in two ways. For fixed frequency and varied angle α they correspond to horizontal lines in Fig. 6, for fixed angle α and varied frequency one gets nearly vertical lines to the right of A which are given by (4.2). Conventional reflection spectra with $n_M = 1$ in (4.2) and fixed angle α are related to straight lines to the left of A in Fig. 6. In Fig. 6 the basic cases of weak and strong anisotropy in uniaxial crystals are given.

It is well known that the dispersion curves of phonon-polaritons are different for spatial and temporal damping. By spatial damping in the regions of TO-LO-splittings the dispersion branches are bent back and k_{BP} is limited by $k_{min} < k_{BP} < k_{max}$ [15]. In an ATR experiment, $1 - R_{TM}(\alpha, \omega)$ may be considered as a response of the sample. This response contains information on the spatial damping of the polaritons, when ω is constant and α is scanned. When α is fixed and ω is scanned, it contains information on mixed temporal and spatial dispersion [65].

Considering Fig. 6 we note that for a polyatomic crystal a great variety of spectra depending on α can be expected. They have to be consistent for different orientations of a crystal. In the actual situation where the optical phonons of very few polyatomic crystals are completely assigned, the ATR-method for TM-polarized radiation may be a helpful tool.

For completeness we give the reflection formula for TE-polarized radiation with respect to Fig. 5

$$R_{TE}(\alpha, \omega) = \left| \frac{(\omega/c) n_M \cos\alpha - [(k_{BP}^{TE})^2 - k_x^2]^{1/2}}{(\omega/c) n_M \cos\alpha = [(k_{BP}^{TE})^2 - k_x^2]^{1/2}} \right|^2 \qquad (4.5)$$

with $k_{BP}^{TE} = (\omega/c)(\varepsilon^{TE})^{1/2}$.

k_{BP}^{TE} means the wave vector of the phonon-polaritons which propagate in x-direction with lattice displacements in y-direction. In Fig. 5 $\varepsilon^{TE} = \varepsilon_y$.

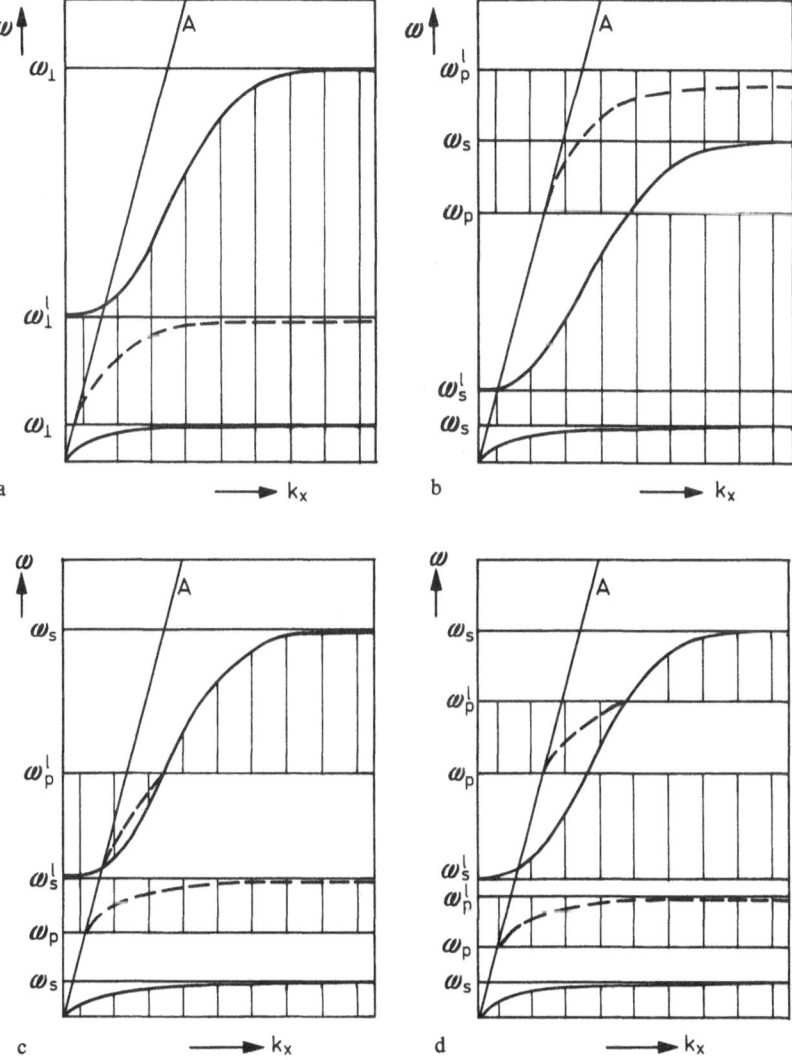

Fig. 6a–d. Display of areas with high reflection for uniaxial crystals [62]. A: dispersion of light in vacuum; hatched areas: high reflection; - - - dispersion of surface phonon-polaritons; horizontal lines: frequencies of phonons; $\omega_{s,p,\perp}$: transverse optical phonons; $\omega^l_{s,p,\perp}$: longitudinal optical phonons; s: lattice displacement perpendicular to the surface; p: lattice displacement parallel to the surface; $\omega^l_s \rightarrow \omega_s$: dispersion of phonon-polaritons; $\perp$ ($\parallel$): lattice displacement perpendicular (parallel) to the optical axis; (a) Ordinary excitations, $\varepsilon_x = \varepsilon_\perp$, $\varepsilon_y = \varepsilon_{\parallel}$, $\varepsilon_z = \varepsilon_\perp$; (b)–(d) Extraordinary excitations ($\varepsilon_x = \varepsilon_y = \varepsilon_\perp$, $\varepsilon_z = \varepsilon_{\parallel}$, or $\varepsilon_y = \varepsilon_z = \varepsilon_\perp$, $\varepsilon_x = \varepsilon_{\parallel}$). (b) Lower reststrahlband (RB) $\omega_s - \omega^l_s$ without overlap of the RB $\omega_p - \omega^l_p$. Upper RB $\omega_s - \omega^l_s$ overlaps higher part of RB $\omega_p - \omega^l_p$. (c) RB $\omega_s - \omega^l_s$ overlaps lower part of RB $\omega_p - \omega^l_p$. (d) Lower RB $\omega_s - \omega^l_s$ overlaps completely lower RB $\omega_p - \omega^l_p$. No overlap of upper RB $\omega_p - \omega^l_p$

Because (4.5) contains only one component of the dielectric tensor, these spectra are not as informative as those for TM-polarization and are less appropriate for the assignment of phonons. They may, however, serve as a proof.

Very often it is not possible to have complete contact between the ATR-crystal and the sample, because both prism and sample are not perfectly plane. Ensuing effects such as decreased reflectivity by surface phonon-polaritons can be taken into account by a multilayer reflection formula for orthorhombic absorbing media derived by Sohler [61]. It is an extension of Wolter's [55] formula and is given for two boundaries by

$$R_{TH} = \left| \frac{(g_2 - g_1)(g_1 + g_0)\exp(\varrho_1 d) + (g_2 + g_1)(g_1 - g_0)\exp(-\varrho_1 d)}{(g_2 + g_1)(g_1 + g_0)\exp(\varrho_1 d) + (g_2 - g_1)(g_1 - g_0)\exp(-\varrho_1 d)} \right|^2 \quad (4.6)$$

with

d: distance between sample and ATR-crystal,
$g_0 = [(\varepsilon_z - n_M^2 \sin^2 \alpha)(\varepsilon_x \varepsilon_z)^{-1/2}]^{1/2}$ with $\mathrm{Re}(g_0) > 0$,
$g_1 = (1 - n_M^2 \sin^2 \alpha)^{1/2}$ with g_1 positive or negative imaginary,
$g_2 = [(1 - \sin^2 \alpha) n_M^{-2}]^{1/2}$ with g_2 positive,
$\varrho_1 = i(\omega/c)(1 - n_M^2 \sin^2 \alpha)^{1/2}$ with ϱ_1 positive or positive imaginary.
$\varepsilon_x, \varepsilon_y, \varepsilon_z$ are related to Fig. 5.

Especially for investigations on surface phonon-polaritons Eq. (4.6) is of great importance, as will be shown in Section 5.

In the Eqs. (4.1), (4.5), and (4.6) for uniaxial crystals damping is included by the extended Kurosawa-formulae [66]

$$\varepsilon_\perp(\omega) = \varepsilon_\perp^\infty \frac{\displaystyle\prod_{i=1}^{m_\perp} (\omega - \omega_{\perp i}^{LO})[\omega + (\omega_{\perp i}^{LO})^*]}{\displaystyle\prod_{i=1}^{m_\perp} (\omega - \omega_{\perp i}^{TO})[\omega + (\omega_{\perp i}^{TO})^*]}$$

and

$$\varepsilon_{||}(\omega) = \varepsilon_{||}^\infty \frac{\displaystyle\prod_{k=1}^{m_{||}} (\omega - \omega_{||k}^{LO})[\omega + (\omega_{||k}^{LO})^*]}{\displaystyle\prod_{k=1}^{m_{||}} (\omega - \omega_{||k}^{TO})[\omega + (\omega_{||k}^{TO})^*]}$$

with $\omega_{v,w}^{(TO,LO)} = \mathrm{Re}(\omega_{v,w}^{(TO,LO)}) - i \cdot \gamma_{v,w}^{(TO,LO)}/2$.

$\gamma_{v,w}^{TO}$ and $\gamma_{v,w}^{LO}$ are the damping factors of the transverse and longitudinal optical phonons. One has $v = \perp, ||$ with $w = 1, ..., m_\perp$ if $v = \perp$, and $w = 1, ..., m_{||}$ if $v = ||$. $m_\perp$ and $m_{||}$ are the numbers of the optical phonons with lattice displacements perpendicular or parallel to the optical axis.

4.2. Experimental Observation of TM-Reststrahlbands

For the measurement of ATR-spectra commercial IR-spectrometers equipped with suitably designed ATR-units [67] and wire grid polarizers on different substrates can be used. Very practical are ATR-crystals with the shape of a hemicylinder as shown in Fig. 7. The beam divergence of the spectrometer, which may be limited by a diaphragm, is further reduced by the focussing effect of the hemicylinder. With our spectrometer, at 500 cm^{-1} one gets an uncertainty for k of $\Delta k < 0.01 \times 10^4$ cm^{-1}. The mirrors M3 and M4 can be turned together with the hemicylinder whereas M1 and M2 are fixed. In this way the angle of inci-

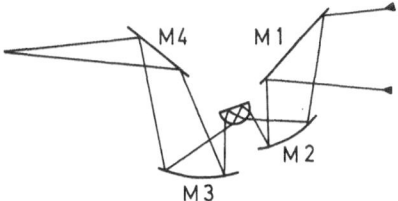

Fig. 7. ATR-unit with hemicylinder

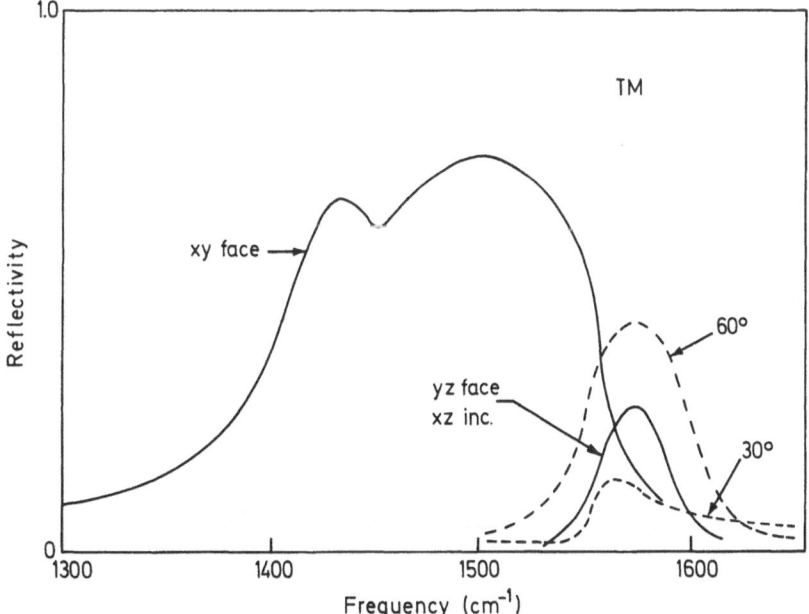

Fig. 8. Reflectivity of two faces of calcite at 37° incidence, v_3 region (TM-polarization). The broken curves are calculated spectra [60]

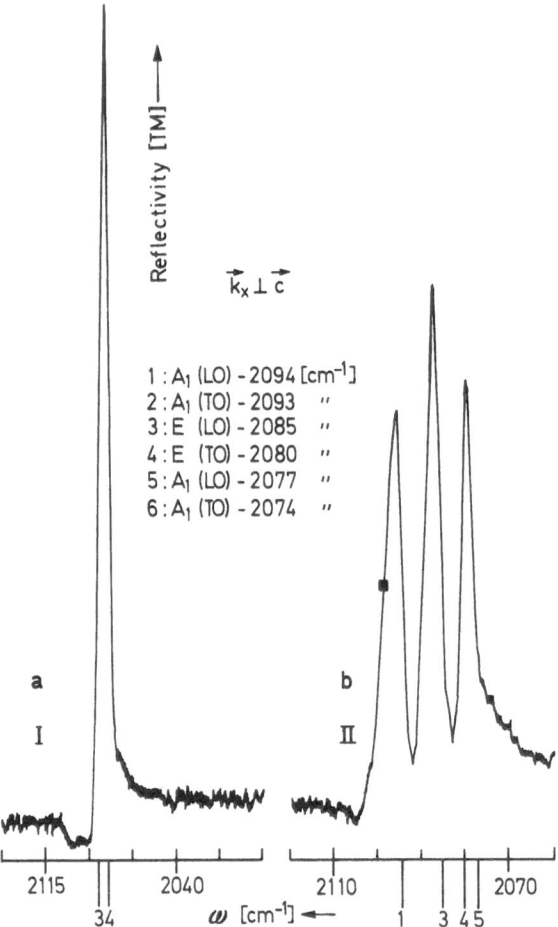

Fig. 9a and b. Reflection and ATR-spectrum of $K_3Cu(CN)_4$ [69]. c: optical axis; (a) reflection spectrum ($n_M = 1$) for $\alpha = 24°$; (b) ATR-spectrum (KRS 5) for $\alpha = 32°$

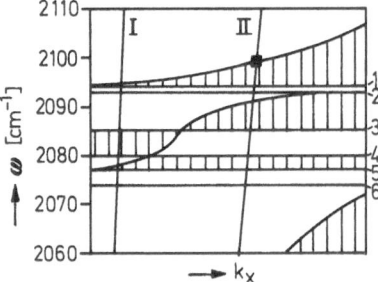

Fig. 10. Display of areas with high reflection for $K_3Cu(CN)_4$ [69]. The dispersion of the extraordinary phonon-polaritons is given qualitatively, ‖‖: see Fig. 6

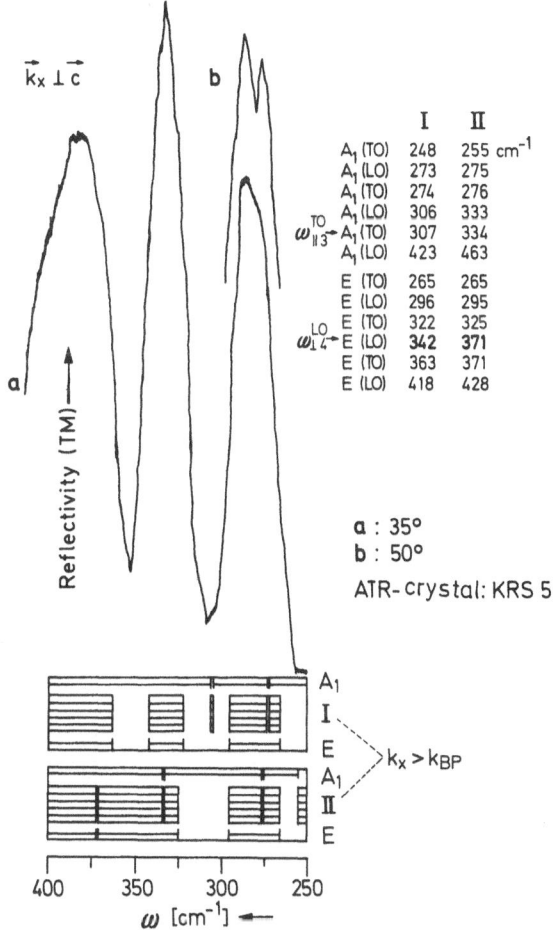

		I	II
	A_1 (TO)	248	255 cm^{-1}
	A_1 (LO)	273	275
	A_1 (TO)	274	276
	A_1 (LO)	306	333
$\omega_{\parallel 3}^{TO}$	A_1 (TO)	307	334
	A_1 (LO)	423	463
	E (TO)	265	265
	E (LO)	296	295
	E (TO)	322	325
$\omega_{\perp 4}^{LO}$	E (LO)	**342**	**371**
	E (TO)	363	371
	E (LO)	418	428

a : 35°
b : 50°
ATR-crystal: KRS 5

Fig. 11. ATR-spectra of lithiumniobate. ⊢⊣: TO-LO-splitting; ≡ : ranges of high reflection for $k > k_{BP}$; I: assignment due to [72]; II: assignment due to [73]

dence α usually may be varied between 20° and 70°. As ATR-hemicylinders mainly serve crystals of AgCl, KRS5 or Ge. They are chosen with regard to their index of refraction. These values are given in [68].

The arrangement of Fig. 7 may also be used for the measurement of reflection spectra with $n_M = 1$. Corresponding spectra for air as medium M are shown in Fig. 8. Decius et al. [60] obtained in this experiment the frequency of the longitudinal optical phonon for the ν_3 band of the carbonate ion in calcite at 1550 cm^{-1}.

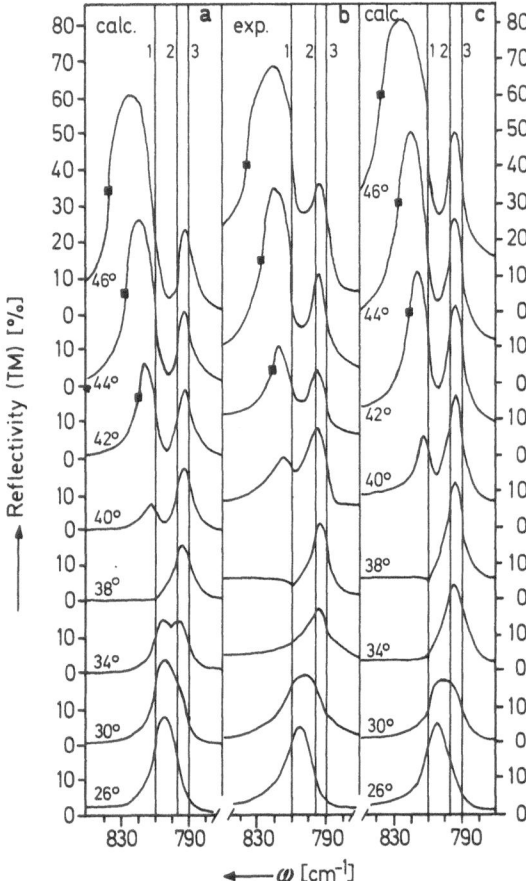

Fig. 12. Measured and calculated reflectivity (TM) of α-quartz [62]. a: calculated spectra (4.3) for spacing $d = 0$; b: measured spectra with $d < 0.25\,\mu\mathrm{m}$ (hemicylinder: KRS 5); c: calculated spectra (4.6) for spacing $d = 0.25\,\mu\mathrm{m}$; 1: $\omega_{\perp 6}^{LO}$; 2: $\omega_{\perp 6}^{TO}$; 3: $\omega_{\parallel 3}^{LO}$; ■: turning points which give the dispersion $\omega_{\parallel 3}^{LO} \rightarrow \omega_{\parallel 4}^{TO}$; $k_x \perp c$

A similar spectrum (I: $n_M = 1$) of $K_3Cu(CN)_4$ in the CN-region is represented in Fig. 9 [69]. Spectrum II is measured by ATR. In Fig. 10 the expected ranges of high reflectivity are indicated by hatched areas. The usefulness of the ATR-spectrum is evident. For example the upper-most longitudinal optical phonon of species A_1 could not be separated from the associated transverse phonon by Raman scattering [69] be-cause the TO-LO-splitting does not exceed $1\,\mathrm{cm}^{-1}$. In a transmission experiment [70] the corresponding absorption is nearly hidden by the adjacent phonon of species E. In the ATR-spectrum the considered

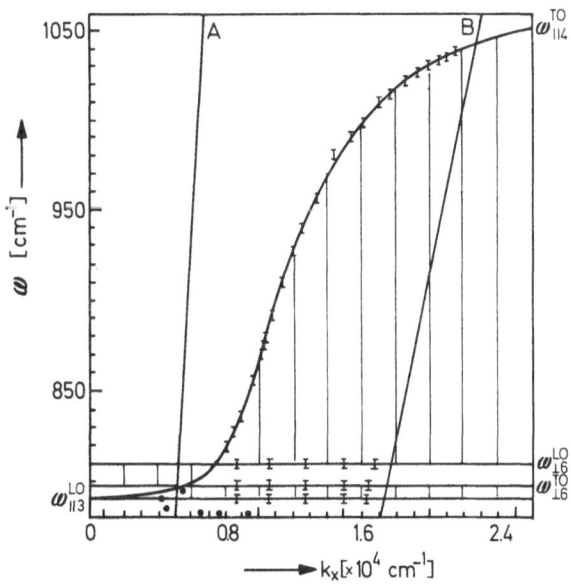

Fig. 13. Dispersion of the extraordinary phonon-polariton $\omega_{\parallel 3}^{LO} \to \omega_{\parallel 4}^{TO}$ and frequencies of the phonons $\omega_{\parallel 3}^{LO}$, $\omega_{\perp 6}^{TO}$, $\omega_{\perp 6}^{LO}$ [62]. A: dispersion of light in vacuum; B: limit of k_x for a Ge-hemicylinder; I: measured values; hatched areas: high reflection; •••: dispersion with spatial damping; curve: dispersion without damping

$A_1(LO)$-phonon determines the right edge of the highest TM-reststrahl-bands (1 in Fig. 9b). The left edge of this band (■) allows the investigation of the phonon-polariton branch which starts at the uppermost $A_1(LO)$-phonon and approaches asymptotically the dispersion line of photons propagating perpendicular to the optical axis in $K_3Cu(CN)_4$.

Figure 11 shows a part of the ATR-spectrum of $LiNbO_3$ for transverse magnetic polarized radiation [71]. The hatched areas are given for two different assignments of the optical phonons and for $k_x > k_{BP}$ in (4.3). Column I in Fig. 11 is the assignment due to Barker and Loudon [72] obtained by IR-reflection ($n_M = 1$). The assignment II is that given by Claus et al. [73]. It was obtained by an extensive investigation of the directional dispersion by Raman scattering. In [73] several erroneous earlier assignments could be corrected. But with respect to $\omega_{\perp 4}^{LO}$ the ATR-spectrum favours I. Otherwise the decreased reflection between $342\,\text{cm}^{-1}$ and $371\,\text{cm}^{-1}$ cannot be explained.

Of great importance for the assignment of phonons is the investigation of the dispersion of phonon-polaritons. The dispersion of the phonon-polariton $\omega_{\parallel 3}^{LO} \to \omega_{\parallel 4}^{TO}$ of α-quartz has been measured by Falge

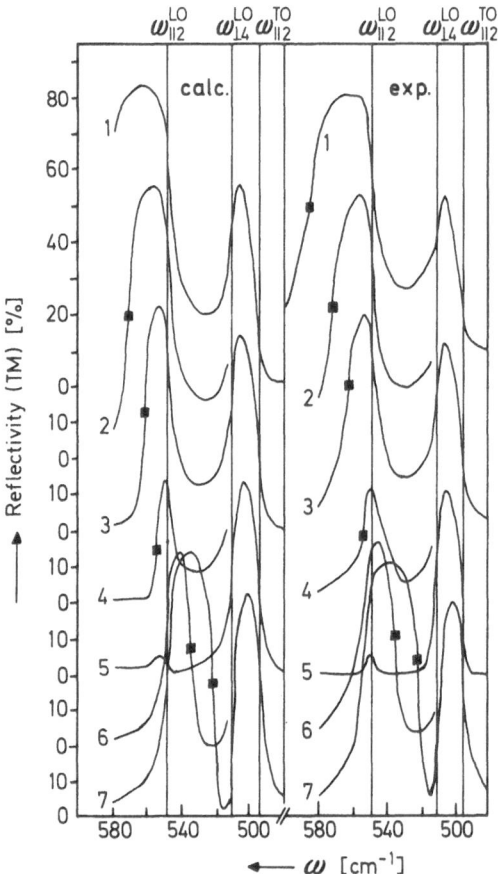

Fig. 14. Measured and calculated reflectivity (TM) of α-quartz [62]. The spectra are cal-culated (4.6) with $d = 0.25\,\mu m$ and measured with $d < 0.25\,\mu m$; ■: turning points which give the dispersion $\omega_{\perp 4}^{LO} \rightarrow \omega_{\perp 5}^{TO}$; hemicylinder: 1–5: KRS5; 6, 7: AgCl; 1: $\alpha = 38°$; 2: $\alpha = 36°$; 3: $\alpha = 34°$; 4: $\alpha = 32°$; 5: $\alpha = 30°$; 6: $\alpha = 32°$; 7: $\alpha = 24°$; $k_x \parallel c$

et al. [62] by ATR for TM-polarized radiation. It cannot be observed by Raman scattering since $\omega_{\parallel 4}^{TO}$ belongs to species A_2. These extra-ordinary phonons of α-quartz are only IR-active. Corresponding cal-culated and measured ATR-spectra are given in Fig. 12. The experi-mental values are compared with the calculated dispersion curve in Fig. 13 and show excellent agreement.

Some of the phonon-polaritons associated with phonons of species E (Raman- and IR-active) already have been measured by Raman scat-tering [74]. The phonon-polariton $\omega_{\perp 4}^{LO} \rightarrow \omega_{\perp 5}^{TO}$ however could not be observed in these experiments because of too low an intensity of $\omega_{\perp 5}^{TO}$.

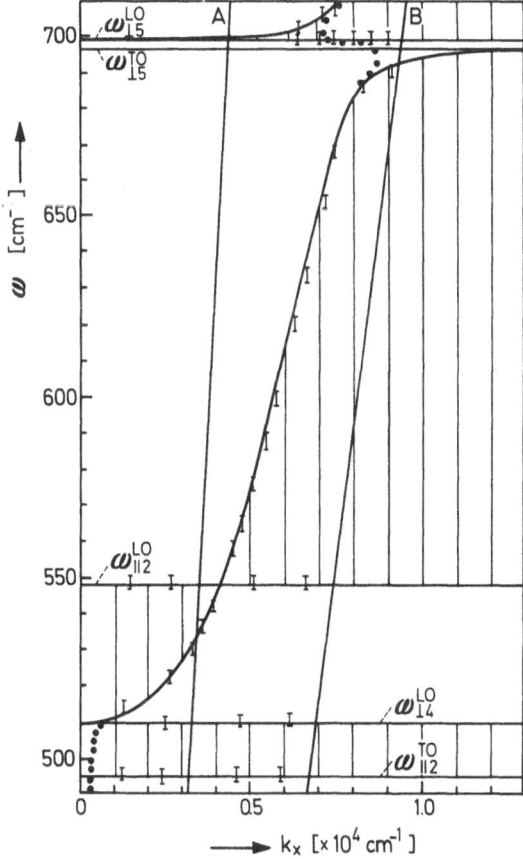

Fig. 15. Dispersion of the ordinary phonon-polariton $\omega_{14}^{LO} \rightarrow \omega_{15}^{TO}$ and frequencies of the phonons $\omega_{\|2}^{TO}$, ω_{14}^{LO}, $\omega_{\|2}^{LO}$ [62]; hatched areas, A, •••, curve, I: see Fig. 13; B: limit of k_x for a KRS 5-hemicylinder

Measurements by ATR are shown in Figs. 14 and 15 [62]. In these experiments the angle of incidence had been fixed and the spectra were recorded by varying the frequency. Calculated spectra (4.6) for fixed frequency and varied angle α are shown in Fig. 16 [75]. In the left insert, the signs ■ mark the turning points of corresponding measured bands. Their projection into the (ω, k) plane agrees with the calculated dispersion curve for spatial damping. Similar results for surface plasmon-polaritons have been reported in [76, 77]. An extensive treatment for spatial damping including bibliography is given in [78].

In addition Figs. 12 and 14 provide the data of optical phonons, some of which are only IR-active.

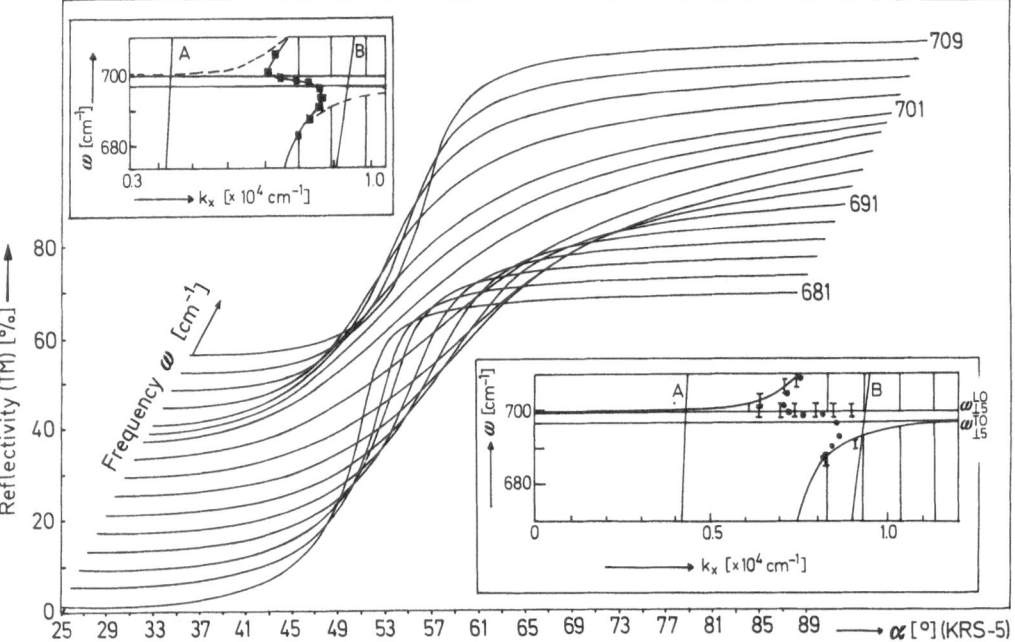

Fig 16. Calculated reflectivity (TM) and measured dispersion at spatial damping of α-quartz [75]. Right insert: see Fig. 15; left insert: ■: measured turning points at fixed frequency and varied angle; input data: [66]; $k_x \parallel c$

5. Optical Excitation of Surface Waves

1968 Otto [2] introduced the ATR-method for the excitation of surface plasmon-polaritons. Meanwhile it turned out to be also suitable for the investigation of surface phonon-polaritons at plane surfaces. In 1972 the first observation of phonon-polaritons propagating at the boundary of a semiinfinite crystal was reported by Marschall and Fischer [79]. An increasing number of publications on ordinary [79–85] and extra-ordinary [62, 86–89] surface phonon-polaritons followed. For a review on the general application of the ATR-method on surface polaritons the reader is referred to [90]. The propagation distances of surface excitations are considered in [91, 92].

The principles of the modified ATR-experiment may be seen in Figs. 17 and 18. In Fig. 17 a TM-polarized ray is incident in the $x - z$ plane under the condition of total reflection $n_1 \sin \alpha > n_M$. The isotropic medium with refractive index n_1 is realized in the experiment by the ATR-hemicylinder as in Fig. 7. At total reflection behind the hemi-

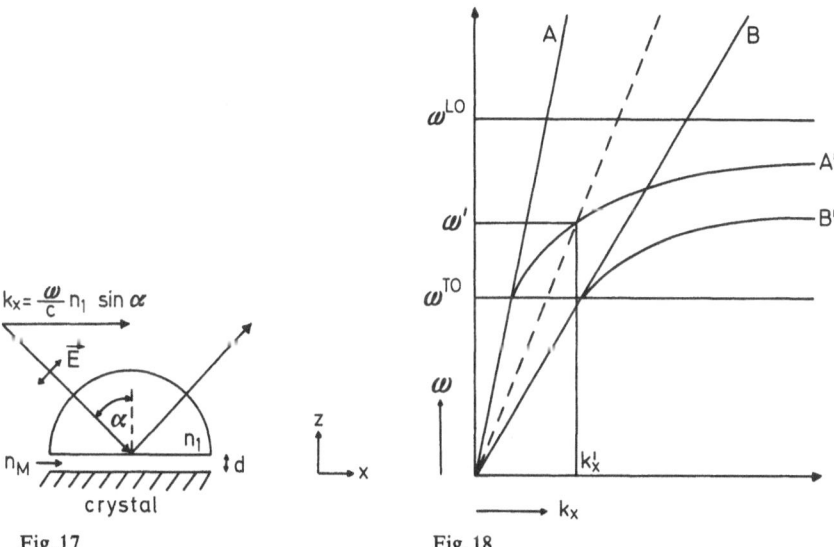

Fig. 17

Fig. 18

Fig. 17. Geometry for the excitation of electromagnetic surface waves

Fig. 18. Dispersion of surface phonon-polaritons. A: dispersion of light in vacuum (air); B: dispersion of light in an isotropic medium with refractive index n_1; A': dispersion of surface phonon-polaritons at the boundary air/sample. B': dispersion of surface phonon-polaritons at the boundary medium n_1/sample; (ω', k_x'): see text; broken line: $k_x = (\omega/c) n_1 \sin\alpha$

cylinder there exists a TM-polarized evanescent wave [67]. The wave vector of this evanescent wave in x-direction is $k_x = (\omega/c) n_1(\omega) \sin\alpha$ and may be varied between the Lines A and B in Fig. 18 depending on α. The sample and the hemicylinder are separated by an isotropic layer ($1 \leq n_M$) which generally consists of air. In this case the gap is provided by foils with thicknesses d from about 1 μm to 30 μm. If the layer is air, the surface phonon-polariton propagates at the boundary air/sample and its dispersion curve starts at the Line A (Fig. 18). In the point (ω', k_x') the phase velocities of the evanescent wave in x-direction and of the surface phonon-polariton are equal and resonance occurs. Then the total reflection is attenuated.

For the investigation of extraordinary surface phonon-polaritons on uniaxial crystals, in Fig. 17 the optical axis has to coincide with x or z. Then Curve C (Fig. 4) corresponds to the dispersion of phonon-polaritons and the dispersion curve of surface phonon-polaritons A' starts at the frequency ω^{TO} or at point P (depending on the position of the reststrahlband parallel and perpendicular to the optical axis).

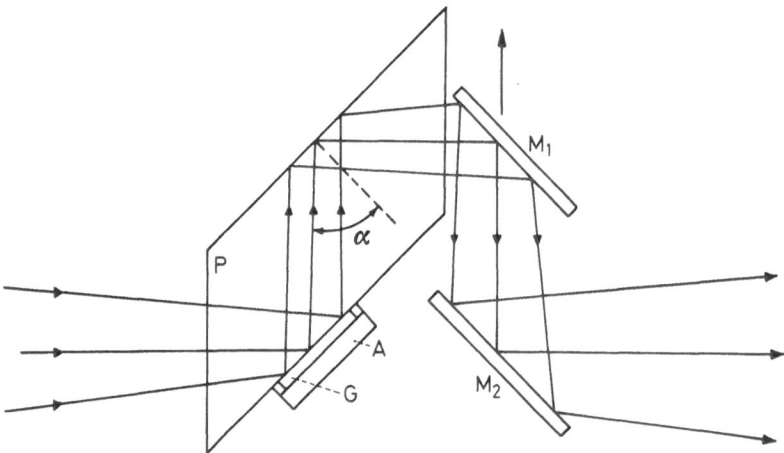

Fig. 19. Representative drawing of a prism P situated in an interferometer with gap G determined by spacers in front of an absorbing sample A [93]. The plane mirror M_1 can be translated while the plane mirror M_2 is fixed

In reflection experiments with $n_1 = n_M = 1$ or with direct contact of the sample and hemicylinder $(n_1 = n_M > 1)$ no excitation of surface waves is possible. In the first case k_x may be varied only at the left side of A and in the second case the dispersion curve of the surface phonon-polaritons (B′) is shifted to the right of B whereas k_x only is variable between A and B. The shift is due to the fact, that the dispersion of surface phonon-polaritons depends on the optical properties of both media at the boundary. Since surface phonon-polaritons only propagate in the frequency regions of TO-LO-splittings the sample is called surface active and the layer surface inactive medium.

Another design of the ATR-crystal is described in [93]. It is shown in Fig. 19. In addition, other shapes of the ATR-crystal may be used [67].

Since we restrict ourselves to the ATR-method and surface waves at the plane surface of a semiinfinite crystal, in the following other investigations for surface phonon-polaritons are only briefly mentioned.

Recently Raman scattering from surface phonon-polaritons in microcrystals has been observed by Scott and Damen [94]. Evans et al. [95] reported Raman scattering from surface phonon-polaritons in a GaAs-film on sapphire. Microcrystals of different shape have been investigated in IR-absorption by Genzel and Martin [96] and Pastrňák [97]. Surface phonon-polaritons in a NaCl-film have been measured by Bryksin et al. [98] using the ATR-method. The excitation of polaritons by the grating technique and the nature of surface plasmon-phonon-polaritons is discussed in [84, 99, 100] and in the literature cited there. A proposition by Bishop and Maradudin [101] to observe surface polaritons by transition radiation has not been realized yet.

5.1. Ordinary Surface Phonon-Polaritons

Nonradiative surface phonons first have been studied by electron scattering experiments in thin films or semiinfinite crystals [102]. Only the effects of film thickness on the surface phonon modes were observed. Optical excitation of nonradiative surface phonon-polaritons which allows detailed investigations has been reported in [79, 103].

Ordinary surface phonon-polaritons occur in cubic and in uniaxial crystals. With regard to the latter, the optical axis is required to be parallel to y (Fig. 3). The dispersion curves are calculated in different ways depending on the phonon damping. An example of weak and frequency independent damping is GaP. Marschall and Fischer [79] got an excellent agreement for the comparison of measured values with a temporal dispersion curve calculated by the relation (3.15) (Fig. 20). This method fails for some alkali-halides. As is well known from reflection spectra these crystals show strong anharmonic contributions to the phonon damping [4] and the corresponding reststrahlbands cannot be

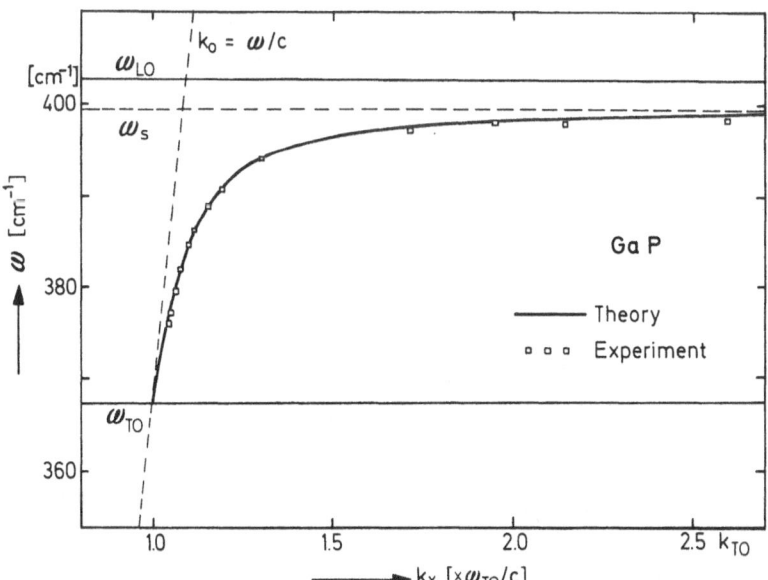

Fig. 20. Dispersion of surface phonon-polaritons in GaP [79]. The theoretical curve is calculated without damping

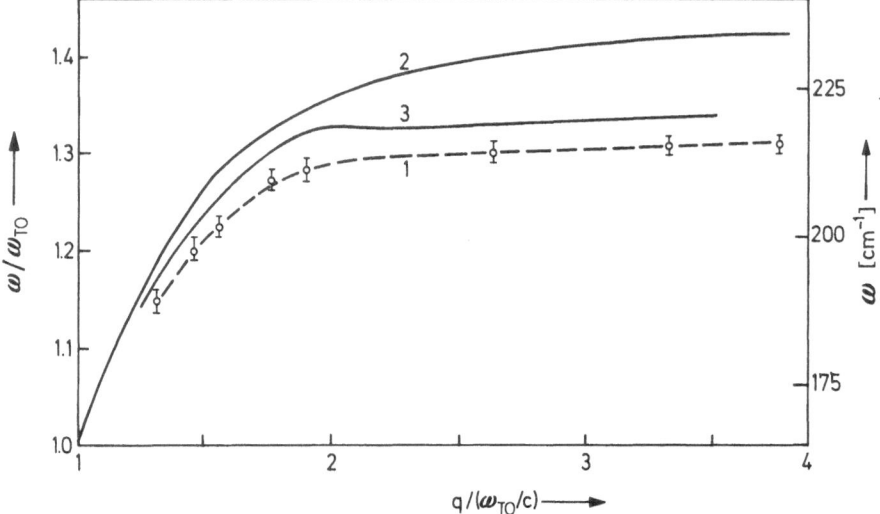

Fig. 21. Dispersion of surface phonon-polaritons in NaCl [81]. 1: experimental curve;
2: calculated in the harmonic approximation; 3: calculated with a second pole at 247 cm^{-1}

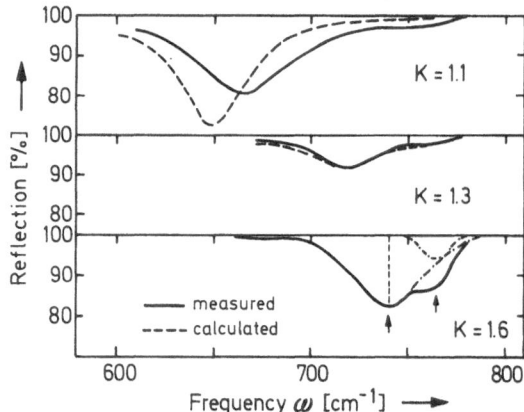

Fig. 22. ATR-spectra of KTaO$_3$ for various wave vectors K, where K is given in multiples
of ω/c_{vac} [85]. Dashed lines: calculated spectra

fitted by using a constant damping parameter. Therefore Bryksin et al.
[81] calculated the ATR-spectra of NaCl for different values of the wave
vector and plotted the minima of the absorption bands (e.g. the minima
of reflection) as dispersion curve. The anharmonic terms are included
by considering a two-oscillator model for the dielectric function. The

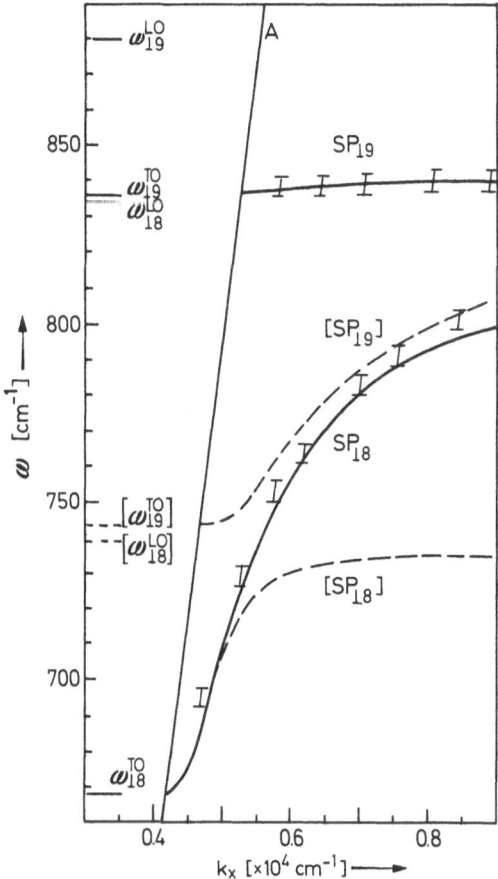

Fig. 23. Dispersion of surface phonon-polaritons in LiNbO$_3$ [71]. A: see Fig. 18

resulting curve in Fig. 21 fits the experimental curve much better. In comparison with other alkali halides the authors found out that the deviation of Curve 2 (Fig. 21) strongly depends on the position of the second pole. If its frequency is greater than the frequency of the surface phonon, as it is the case for NaCl, the harmonic approximation leads to too high a value of the surface vibration. Furthermore the calculation of ATR-spectra with frequency dependent damping explains asymmetries of the bands. Corresponding spectra are shown for KTaO$_3$ in Fig. 22. Fischer et al. [85] interpret the second minimum due to an additive combination band. The calculated spectra are based on a dielectric function obtained from a Kramers-Kronig analysis of normal incidence reflectivity which shows fine structure in this region. The

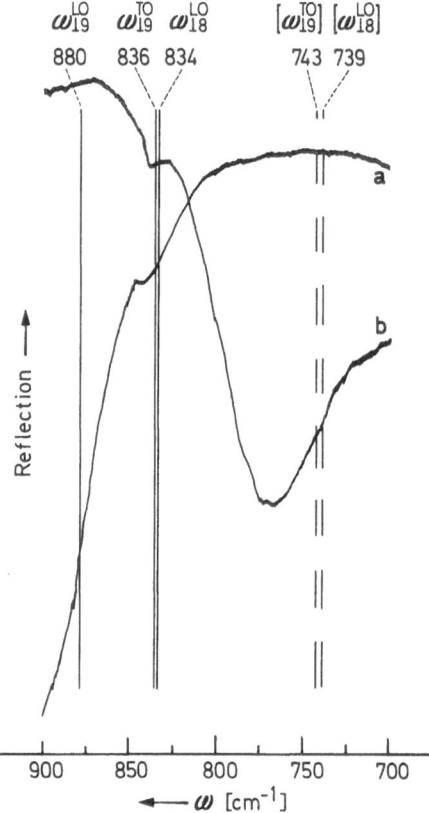

Fig. 24a and b. TE-reststrahlbands, (a) and ATR-spectrum, (b) for LiNbO$_3$ [71]. (a) $\alpha = 24°$; the optical axis is perpendicular to the surface; (b) ATR-hemicylinder: KRS 5; $\alpha = 35°$; $d = 0.7\,\mu m$; the optical axis is parallel to y in Fig. 3

dispersion of the two bands could theoretically be fitted by a response function derived by Barker [104].

Besides the investigation of phonon damping surface phonon-polaritons may be used for the assignment of optical phonons.

In [73] for LiNbO$_3$ the frequencies of $\omega_{\perp 9}^{TO}$ and $\omega_{\perp 8}^{LO}$ had been chosen from some bands of low Raman intensity. This assignment leads to the dispersion curves of surface phonon-polaritons which are given in Fig. 23 by dashed lines. Since [SP$_{\perp 8}$] cannot be observed experimentally and the measured dispersion curve of SP$_{\perp 8}$ passes the frequencies of [$\omega_{\perp 8}^{LO}$] and [$\omega_{\perp 9}^{TO}$] unperturbed the Raman band at 740 cm^{-1} must be due to a two-phonon process. A new assignment of $\omega_{\perp 8}^{LO}$ and $\omega_{\perp 9}^{TO}$ is proposed in Fig. 24, where the ATR-spectrum (b) shows a small second band. The fine structure in the reflection spectrum (a) which had not been reported in [72] indicates the missing phonons for LiNbO₃. Further work on LiNbO₃ for a complete reassignment of the phonon structure is in progress.

As assignment of ordinary surface phonon-polaritons (SP) it is proposed in [87] to characterize them by the number of the reststrahlband in which they occur. The reststrahlbands are numbered from low to high frequencies. The index $_\perp$ refers to $\varepsilon_\perp$.

5.2. Extraordinary Surface Phonon-Polaritons

In [87] for regular extraordinary surface phonon-polaritons (e.o. SP) the following assignment is proposed. The indices $\| i$, $\perp i$ of e.o. $SP_{\|i, \perp i}$ depend on $\varepsilon_\|$ and $\varepsilon_\perp$ respectively and on the number of the reststrahlbands. Virtual and real excitation surface phonon-polaritons are distinguished by the sign ∞ for real excitation (e.o. SP^∞), because these polaritons still exist for $k_x \to \infty$. The frequency of the corresponding surface phonon is called ω_{sij} since it is given by the reststrahlbands in $\varepsilon_\perp$ and in $\varepsilon_\|$ together. If these assignments are not unambiguous, the frequency of the surface phonon ω_{sij} is added to e.o. SP^∞. This holds for uniaxial crystals with the optical axis being parallel to k_x or perpendicular to k_x and the surface. With regard to biaxial crystals the orientation of the principal axes has to be parallel to $\hat{x}$, $\hat{y}$, $\hat{z}$ (Fig. 1) and $_\|$, $_\perp$ have to be changed by x, y, z depending on ε_x, ε_y, ε_z. Surface phonon-polaritons in biaxial crystals, however, have not been investigated yet.

Table 1. Ranges of ϑ for the non regular TM-polarized surface phonon-polaritons of α-quartz [105]

$\vartheta = 0°$	$\alpha < \vartheta < \beta$	$\vartheta = 90°$
e.o. $SP_{\perp 1}$ ———	e.o. SP_1 $(0° - 62°)$	
e.o. $SP_{\perp 2}$ ———	e.o. SP_2 $(0° - 65°)$	
	e.o. SP_3 $(15° - 90°)$ ———	e.o. $SP_{\|1}$
e.o. $SP_{\perp 3}$ ———	e.o. SP_4 $(0° - 47°)$	
e.o. $SP^\infty_{\perp 4}$ ———	e.o. SP^∞_5 $(0° - 90°)$ ———	e.o. $SP^\infty_{\|2}$
	e.o. SP_6 $(58° - 90°)$ ———	e.o. $SP_{\|2}$
e.o. $SP_{\perp 5}$ ———	e.o. SP_7 $(0° - 55°)$	
	e.o. SP_8 $(20° - 90°)$ ———	e.o. $SP_{\|3}$
e.o. $SP_{\perp 6}$ ———	e.o. SP_9 $(0° - 47°)$	
e.o. $SP^\infty_{\perp 7}$ ———	e.o. SP^∞_{10} $(0° - 90°)$ ———	e.o. $SP^\infty_{\|4}$ (ω_{s47})
e.o. $SP^\infty_{\perp 8}$ ———	e.o. SP^∞_{11} $(0° - 90°)$ ———	e.o. $SP^\infty_{\|4}$ (ω_{s48})

α and β are the limiting angles for $\vartheta \cdot \vartheta$: see Fig. 3.

Non regular TM-polarized surface phonon-polaritons exist in the geometry shown in Fig. 3 with $0 < \vartheta < \pi/2$. Their dispersion branches are numbered with increasing frequencies. The system of extraordinary surface phonon-polaritons in α-quartz may serve as an example which is given in Table 1 [105]. In addition Table 1 contains the limiting ranges of non regular TM-polarized virtual excitation surface phonon-polaritons. Since α-quartz is an uniaxial polyatomic crystal which allows all theoretically possible effects of extraordinary surface phonon-polaritons to be studied the frequencies of its phonons are given in Table 2. From these data the expected branches of surface phonon-polaritons may easily be evaluated as shown qualitatively for an interval of ω in Fig. 25. The calculated dispersion curves are contained in Figs. 26 and 27. In spite of neglecting damping, they agree with the measured values. All theoretical predictions concerning stop points of virtual excitation surface phonon-polaritons, stop band for $0 < \varepsilon_s < 1$ (Fig. 6d) and limiting frequencies of real excitation surface phonon-polaritons are fulfilled. The dispersion branch of e.o. $SP_{\perp 4}^{\infty}$ shows no influence of the change from virtual to real excitation surface phonon-polariton [87].

The existence of stop points is of considerable interest since the vanishing of the corresponding bands allows for a direct measurement of longitudinal optical phonons. Spectra of e.o. $SP_{\parallel 2}$ giving the frequency of $\omega_{\parallel 2}^{LO}$ are shown in Fig. 28 for wave vectors to the right and the left of the stop point. As the spectra in Fig. 28 are measured for small distances between the ATR-hemicylinder and the α-quartz crystal bulk effects already appear. The change for different penetration depths of

Table 2. Phonons of α-quartz [66]

$\varepsilon_{\perp}^{\infty} = 2.356$	
$\omega_{\perp 1}^{TO} = 128.3$ cm^{-1}	$\omega_{\perp 1}^{LO} = 129.0$ cm^{-1}
$\omega_{\perp 2}^{TO} = 265$ cm^{-1}	$\omega_{\perp 2}^{LO} = 270.2$ cm^{-1}
$\omega_{\perp 3}^{TO} = 394$ cm^{-1}	$\omega_{\perp 3}^{LO} = 402.7$ cm^{-1}
$\omega_{\perp 4}^{TO} = 450$ cm^{-1}	$\omega_{\perp 4}^{LO} = 509.8$ cm^{-1}
$\omega_{\perp 5}^{TO} = 697$ cm^{-1}	$\omega_{\perp 5}^{LO}$ 699.0 cm^{-1}
$\omega_{\perp 6}^{TO} = 797$ cm^{-1}	$\omega_{\perp 6}^{LO} = 809.7$ cm^{-1}
$\omega_{\perp 7}^{TO} = 1072$ cm^{-1}	$\omega_{\perp 7}^{LO} = 1160$ cm^{-1}
$\omega_{\perp 8}^{TO} = 1163$ cm^{-1}	$\omega_{\perp 8}^{LO} = 1237$ cm^{-1}

$\varepsilon_{\parallel}^{\infty} = 2.383$	
$\omega_{\parallel 1}^{TO} = 364$ cm^{-1}	$\omega_{\parallel 1}^{LO} = 387.9$ cm^{-1}
$\omega_{\parallel 2}^{TO} = 495$ cm^{-1}	$\omega_{\parallel 2}^{LO} = 547.7$ cm^{-1}
$\omega_{\parallel 3}^{TO} = 778$ cm^{-1}	$\omega_{\parallel 3}^{LO} = 789.8$ cm^{-1}
$\omega_{\parallel 4}^{TO} = 1080$ cm^{-1}	$\omega_{\parallel 4}^{LO} = 1239$ cm^{-1}

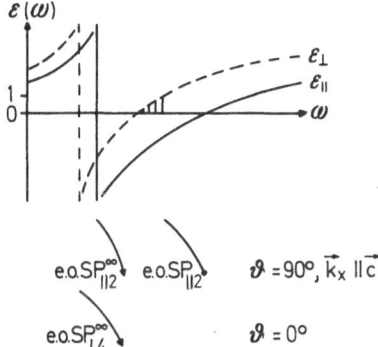

Fig. 25. Qualitative curve of $\varepsilon_\perp$ and $\varepsilon_\parallel$ for α-quartz at $\omega_{\parallel 2}^{TO} < \omega_{\perp 4}^{LO} < \omega_{\parallel 2}^{LO}$ and expected dispersion branches of surface phonon-polaritons (related to the ω-axis). ϑ: see Fig. 3; c: optical axis; $\rightarrow$: real excitation; $\longrightarrow$: virtual excitation; $\|\|$: stop band for $\vartheta = 90°$ and $k_x \| c$

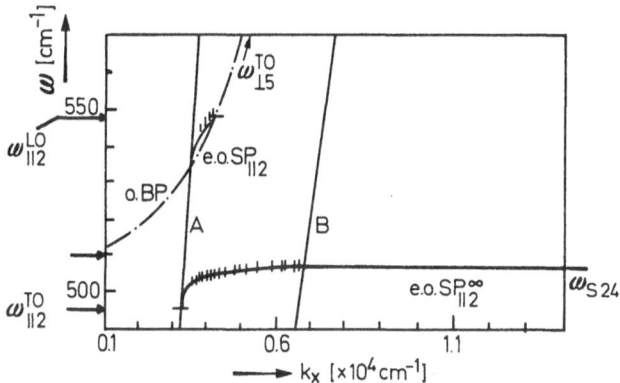

Fig. 26. Dispersion of e.o. $SP_{\parallel 2}$ and e.o. $SP_{\parallel 2}^\infty$ of α-quartz [87]. A, B: see Fig. 15; $-\cdot-\cdot-$: phonon-polariton $\omega_{\perp 4}^{LO} \to \omega_{\perp 5}^{TO}$; $\Delta k_x < 0.01 \times 10^4$ cm^{-1} at 500 cm^{-1}

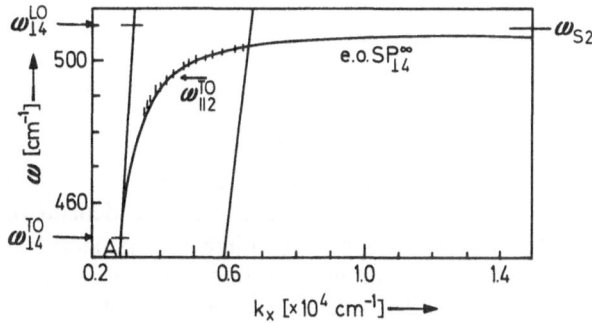

Fig. 27. Dispersion of e.o. $SP_{\perp 4}^\infty$ of α-quartz [87]. A, B: see Fig. 15; $\rightarrow$: change from virtual to real excitation surface phonon-polariton; $\Delta k_x < 0.01 \times 10^4$ cm^{-1} at 500 cm^{-1}

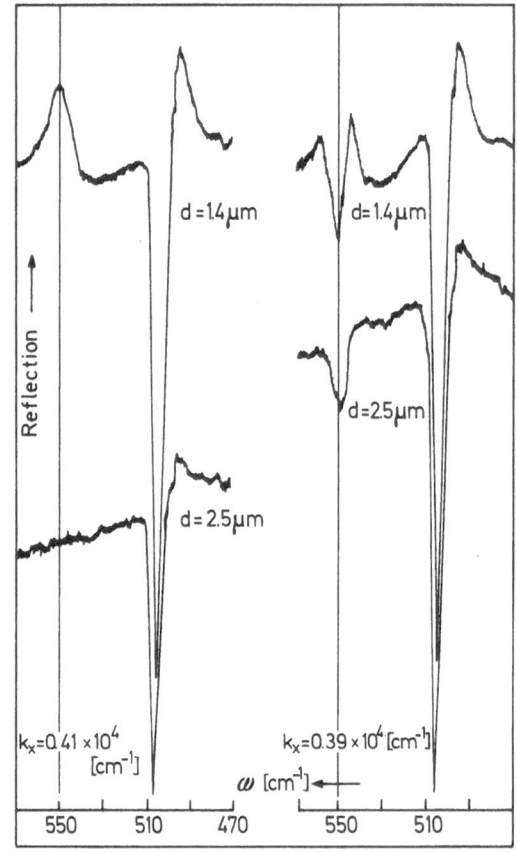

Fig. 28. The stop point of the virtual excitation surface phonon-polariton e.o. SP$_{||2}$ (α-quartz) near 550 cm^{-1} for two air gaps of thickness d and two values of k for $\omega = 550$ cm^{-1}, together with the real excitation surface phonon-polariton SP$_{||2}^{\infty}$ near 510 cm^{-1} [87]

the evanescent wave from TM-reststrahlbands at $d = 0$ μm to total reflection is described in Fig. 29. These spectra may be compared with Fig. 14.

From the very specific behaviour of the dispersion of surface phonon-polaritons, one can conclude that one has rather a sensitive method for the assignment of phonons as all effects strongly depend on the frequencies of phonons. There are very different dispersion branches for weak or strong anisotropy which in addition have to be consistent for different orientations of the investigated crystal. The basic sequences of ω^{TO} and ω^{LO} and the resulting dispersion branches of surface phonon-polaritons are included in Fig. 6. In more complicated cases as for the

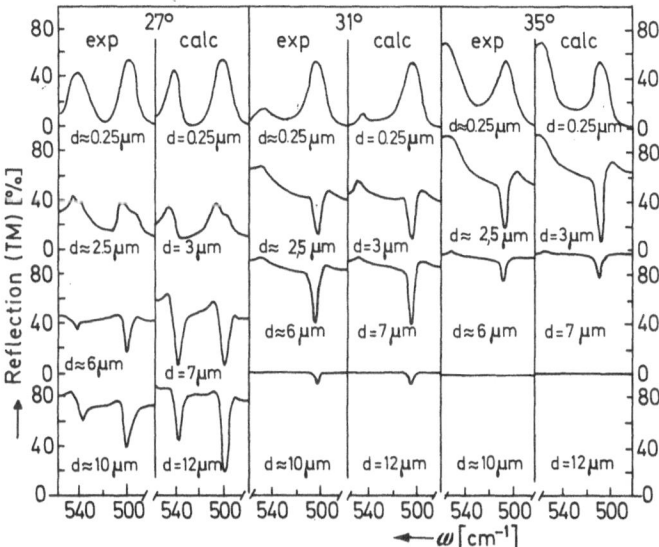

Fig. 29. Measured and calculated TM-reflection in the interval of e.o. $SP_{\|2}$ and e.o. $SP_{\|2}^{\infty}$ (α-quartz) [62]. For a comparison see Figs. 6d, 14, 15, 26, and 28

region 1000 cm^{-1} to 1200 cm^{-1} of α-quartz they can be combined in a straightforward way.

The results for α-quartz in [87] are confirmed in [89]. Another study of extraordinary surface phonon-polaritons has been reported in [86] for MgF$_2$ and TiO$_2$. As a test of optical constants of CdS, extraordinary surface phonon-polaritons have been used by Perry et al. [88]. In all these publications regular extraordinary polaritons are discussed. Non regular extraordinary ones are treated for α-quartz in [105].

Finally it is to be mentioned that Dubovskii [106] first postulated stop points for the branches of virtual excitation surface polaritons. Some unretarded solutions of real excitation surface polaritons had already been given by Kulik [107] and by Agranovich and Dubovskii [108].

Acknowledgements. We thank Prof. Dr. J. Brandmüller and Prof. Dr. L. Merten for their interest in this work. For the financial support we are very much obliged to the Deutsche Forschungsgemeinschaft.

References

1. Brandmüller, J., Claus, R., Merten, L.: Springer Tracts Mod. Phys. (to be published).
2. Otto, A.: Z. Physik **216**, 398 (1968).
3. Huang, K.: Proc. Roy. Soc. (London) A **208**, 352 (1951)

4. Born, M., Huang, K.: Dynamical theory of crystal lattices. Oxford: Clarendon Press 1954.
5. Barker, A. S.: Phys. Rev. **136**, 1290 (1964).
6. Loudon, R.: Proc. Phys. Soc. (London) **82**, 393 (1963).
7. Loudon, R.: Adv. Phys. **13**, 423 (1964).
8. Merten, L.: Z. Naturforsch. **17**a, 65 (1961).
9. Merten, L.: Z. Naturforsch. **22**a, 359 (1967).
10. Merten, L.: Phys. Stat. Sol. **30**, 449 (1968).
11. Pick, R.: Adv. Phys. **19**, 269 (1970).
12. Scott, J. F.: Am. J. Phys. **39**, 1360 (1971).
13. Barker, A. S., Loudon, R.: Rev. Mod. Phys. **44**, 18 (1972).
14. Merten, L.: Atomic structure and properties of solids, p. 119. New York: Academic Press 1972.
15. Merten, L.: Advances in solid state physics, XII, p. 343. Madelung, O. (Ed.): Braunschweig: Pergamon Vieweg 1972.
16. Kurosawa, T.: J. Phys. Soc. Japan **16**, 1298 (1961).
17. Lyddane, R. H., Sachs, R. G., Teller, E.: Phys. Rev. **59**, 673 (1941).
18. Cochran, W., Cowley, R. A.: J. Phys. Chem. Solids **23**, 447 (1962).
19. Borstel, G., Merten, L.: Z. Naturforsch. **26**a, 653 (1971).
20. Merten, L., Borstel, G.: Z. Naturforsch. **27**a, 1073 (1972).
21. Unger, B., Schaack, G.: Phys. Stat. Sol. (b) **48**, 285 (1971).
22. Unger, B.: Phys. Stat. Sol. (b) **49**, 107 (1972).
23. Borstel, G., Merten, L.: Z. Naturforsch. **28**a, 1038 (1973).
24. Burstein, E., Ushioda, S., Pinczuk, A., Scott, J. F.: Light scattering of solids, p. 47. Berlin-Heidelberg-New York: Springer 1969.
25. Rosenstock, H. B.: Phys. Rev. **121**, 416 (1961).
26. Maradudin, A. A., Weiss, H. G.: Phys. Rev. **123**, 1968 (1961).
27. Lucas, A. A.: Phys. Rev. **162**, 801 (1967).
28. Fuchs, R., Kliewer, K. L.: Phys. Rev. A **140**, 2076 (1965).
29. Kliewer, K. L., Fuchs, R.: Phys. Rev. **144**, 495 (1966).
30. Kliewer, K. L., Fuchs, R.: Phys. Rev. **150**, 573 (1966).
31. Fuchs, R., Kliewer, K. L.: J. Opt. Soc. Amer. **58**, 319 (1968).
32. Englman, R., Ruppin, R.: J. Phys. C **1**, 614 (1968).
33. Ruppin, R., Englman, R.: J. Phys. C **1**, 630 (1968).
34. Englman, R., Ruppin, R.: J. Phys. C **1**, 1515 (1968).
35. Barron, T. H. K.: Phys. Rev. **123**, 1995 (1961).
36. Ruppin, R., Englman, R.: Rep. Prog. Phys. **33**, 149 (1970).
37. Feuchtwang, R. E.: Phys. Rev. **155**, 715 (1967).
38. Feuchtwang, T. E.: Phys. Rev. **155**, 731 (1967).
39. Lucas, A. A.: J. Chem. Phys. **48**, 3156 (1968).
40. Tong, S. Y., Maradudin, A. A.: Phys. Rev. **181**, 1318 (1969).
41. Martin, T. P.: Phys. Rev. B **7**, 3906 (1973).
42. Trullinger, S. E., Cunningham, S. L.: Phys. Rev. B **8**, 2622 (1973).
43. Lyubimov, V. N., Sannikov, D. G.: Soviet Phys.-Solid State **14**, 575 (1972).
44. Hartstein, A., Burstein, E., Brion, J. J., Wallis, R. F.: Surf. Sci. **34**, 81 (1973).
45. Hartstein, A., Burstein, E., Brion, J. J., Wallis, R. F.: Solid State Commun. **12**, 1083 (1973).
46. Burstein, E., Hartstein, A., Schoenwald, J., Maradudin, A. A., Mills, D. L., Wallis, R. F.: Proc. Taormina Res. Conf. Structure of Matter. Taormina, 1972 (to be published).
47. Borstel, G.: Z. Naturforsch. **28**a, 1055 (1973).
48a. Borstel, G.: Phys. Stat. Sol. (b) **60**, 427 (1973).
48b. Otto, A.: Unpublished comment.

49. Borstel, G.: Doctoral Thesis, University Münster (1974).
50. Fano, U.: Ann. Phys. **32**, 393 (1938).
51. Fano, U.: J. Opt. Soc. Am. **31**, 213 (1941).
52. Drude, P.: Ann. Phys. Chem. N. F. **32**, 619 (1887).
53. Szivessy, G.: Handbuch der Physik, Vol. 20. Flügge, S. (Ed.). Berlin: Springer 1928.
54. Schopper, H.: Z. Physik **132**, 146 (1952).
55. Wolter, H.: Handbuch der Physik, Vol. 24. Flügge, S. (Ed.). Berlin-Göttingen-Heidelberg: Springer 1956.
56. Flournoy, P. A., Schaffers, W. J.: Spectrochim. Acta **22**, 5 (1966).
57. Bell, E. E.: Handbuch der Physik, Vol. 25/2a. Flügge, S. (Ed.). Berlin-Heidelberg-New York: Springer 1967.
58. Mosteller, L. P., Wooten, F.: J. Opt. Soc. Am. **58**, 511 (1968).
59. Koch, E. E., Otto, A., Kliewer, K. L.: Chem. Phys. **3**, 362 (1974).
60. Decius, J. C., Frech, R., Brüsch, P.: J. Chem. Phys. **58**, 4056 (1973).
61. Sohler, W.: Opt. Commun. **10**, 203 (1974).
62. Falge, H. J., Otto, A., Sohler, W.: Phys. Stat. Sol. (b) **63**, 259 (1974).
63. Hisano, K., Okamoto, Y., Matumura, O.: J. Phys. Soc. Japan **28**, 425 (1970).
64a. Barnes, R. B.: Z. Physik **75**, 723 (1932).
64b. Berreman, D. W.: Phys. Rev. **130**, 2193 (1963).
65. Otto, A.: Proc. Taormina Res. Conf. Structure of Matter. Taormina 1972 (to be published).
66. Lamprecht, G., Merten, L.: Phys. Stat. Sol. (b) **55**, 33 (1973).
67. Harrick, N. J.: Internal reflection spectroscopy. New York: J. Wiley & Sons 1967.
68. Smith, R. A., Jones, F. E., Chasmar, R. P.: The detection and measurement of infrared radiation. Oxford: Clarendon Press 1968.
69. Nitsch, W., Falge, H. J., Claus, R.: Z. Naturforsch. **29a**, 1017 (1974).
70. Jones, L. H.: J. Chem. Phys. **29**, 463 (1958).
71. Falge, H. J.: Doctoral Thesis, University München (1974).
72. Barker, A. S., Loudon, R.: Phys. Rev. **158**, 433 (1967).
73. Claus, R., Borstel, G., Wiesendanger, E., Steffan, L.: Z. Naturforsch. **27a**, 1187 (1972).
74. Scott, J. F., Cheesman, L. S., Porto, S. P. S.: Phys. Rev. **162**, 834 (1967).
75. Schuller, E., Falge, H. J., Brandmüller, J.: (To be published).
76. Arakawa, E. T., Williams, M. W., Hamm, R. N., Ritchie, R. H.: Phys. Rev. Letters **31**, 1127 (1973).
77. Alexander, R. W., Kovener, G. S., Bell, R. J.: Phys. Rev. Letters **32**, 154 (1974).
78. Otto, A.: Optik **38**, 566 (1973).
79. Marschall, N., Fischer, B.: Phys. Rev. Letters **28**, 811 (1972).
80. Fischer, B.: Doctoral Thesis, Technische Hochschule Stuttgart (1973).
81. Bryksin, V. V., Gerbshtein, Yu. M., Mirlin, D. N.: Phys. Stat. Sol. **51**, 901 (1972).
82. Barker, A. S.: Phys. Rev. Letters **28**, 892 (1972).
83. Fischer, B., Tyler, I. L., Bell, R. J.: Proc. Taormina 1972 (to be published).
84. Fischer, B., Marschall, N., Queisser, H. J.: Surface Sci. **34**, 50 (1973).
85. Fischer, B., Bäuerle, D., Buckel, W. J.: Solid State Commun. **14**, 291 (1974).
86. Bryksin, V. V., Mirlin, D. N., Reshina, J. J.: JETP Letters **16**, 445 (1972) and Soviet Phys.-Solid State **15**, 760 (1973).
87. Falge, H. J., Otto, A.: Phys. Stat. Sol. (b) **56**, 523 (1973).
88. Perry, C. H., Fischer, B., Buckel, W.: Solid State Commun. **13**, 1261 (1973).
89. Reshina, J. J., Zolotarev, B. M.: Fiz. Tver. Tela. **15**, 3020 (1973).
90. Otto, A.: Advances in solid state physics. Braunschweig: Pergamon Vieweg 1974 (to be published).
91. Schoenwald, J., Burstein, E., Elson, J. M.: Solid State Commun. **12**, 185 (1973).
92. Bell, R. J., Alexander, R. W., Parks, W. F., Kovener, G.: Optics Commun. **8**, 147 (1973).

148   *G. Borstel et al.*

93. Bell, R. J., Fischer, B., Tyler, I. L.: Appl. Optics **12**, 832 (1973).
94. Scott, J. F., Damen, T. C.: Opt. Commun. **5**, 410 (1972).
95. Evans, D. J., Ushioda, S., McMullen, J. D.: Phys. Rev. Letters **31**, 369 (1973).
96. Genzel, L., Martin, T. P.: Phys. Stat. Sol. (b) **51**, 91 (1972).
97. Pastrňák, J.: Phys. Stat. Sol. **3**, 657 (1970).
98. Bryksin, V. V., Gerbshtein, Yu. M., Mirlin, D. N.: Soviet Phys.-Solid State **14**, 2849 (1973).
99. Madelung, O.: Lokalisierte Zustände. Festkörpertheorie, III, Chapter XV, p. 108. Heidelberger Taschenbücher. Berlin-Heidelberg-New York: Springer 1973.
100. Bryksin, V. V., Mirlin, D. N., Reshina, I. I.: Solid State Commun. **11**, 695 (1972).
101. Bishop, M. F., Maradudin, A. A.: Solid State Commun. **12**, 1225 (1973).
102. Boersch, H., Geiger, J., Stickel, W.: Z. Physik **212**, 130 (1968).
103. Bryksin, V. V., Gerbshtein, Yu. M., Mirlin, D. N.: Fiz. Tverd. Tela. **13**, 2125 (1971); Soviet Phys.-Solid State **13**, 1779 (1972).
104. Barker, A. S.: Surface Sci. **34**, 62 (1973).
105. Falge, H. J., Borstel, G., Otto, A.: Phys. Stat. Sol. (b) **65**, 123 (1974).
106. Dubovskii, O. A.: Soviet Phys.-Solid State **12**, 2471 (1971).
107. Kulik, I. O.: Soviet Phys.-JETP **15**, 380 (1962).
108. Arganovich, V. M., Dubovskii, O. A.: Soviet Phys.-Solid State **7**, 2343 (1966).

Dr. G. Borstel
Fachbereich Physik der Universität Münster
Lehrstuhl Prof. Dr. L. Merten
D-4400 Münster
Schloßplatz 7
Federal Republic of Germany

Dr. H. J. Falge
Sektion Physik der Universität München
Lehrstuhl Prof. Dr. J. Brandmüller
D-8000 München 40
Schellingstr. 4/IV
Federal Republic of Germany

Dr. A. Otto
Sektion Physik der Universität München
Lehrstuhl Prof. Dr. W. Rollwagen
New address:
Max-Planck-Institut für Festkörperphysik
D-7000 Stuttgart 1
Heilbronner Str. 69
Federal Republic of Germany

SPRINGER TRACTS IN MODERN PHYSICS

Ergebnisse der exakten Naturwissenschaften

Atomic and Molecular Physics

Dettmann, K.: High Energy Treatment of Atomic Collisions (Vol. 58)

Donner, W., Süßmann, G.: Paramagnetische Felder am Kernort (Vol. 37)

Langbein, D.: Theory of Van der Waals Attraction (Vol. 72)

Racah, G.: Group Theory and Spectroscopy (Vol. 37)

Seiwert, R.: Unelastische Stöße zwischen angeregten und unangeregten Atomen (Vol. 47)

Zu Putlitz, G.: Determination of Nuclear Moments with Optical Double Resonance (Vol. 37)

Elementary Particle Physics

Current Algebra

Furlan, G., Paver, N., Verzegnassi, C.: Low Energy Theorems and Photo- and Electroproduction Near Threshold by Current Algebra (Vol. 62)

Gatto, R.: Cabibbo Angle and $SU_2 \times SU_2$ Breaking (Vol. 53)

Genz, H.: Local Properties of σ-Terms: A Review (Vol. 61)

Kleinert, H.: Baryon Current Solving SU (3) Charge-Current Algebra (Vol. 49)

Leutwyler, H.: Current Algebra and Lightlike Charges (Vol. 50)

Mendes, R. V., Ne'eman, Y.: Representations of the Local Current Algebra. A Constructional Approach (Vol. 60)

Müller, V. F.: Introduction to the Lagrangian Method (Vol. 50)

Pietschmann, H.: Introduction to the Method of Current Algebra (Vol. 50)

Pilkuhn, H.: Coupling Constants from PCAC (Vol. 55)

Pilkuhn, H.: S-Matrix Formulation of Current Algebra (Vol. 50)

Renner, B.: Current Algebra and Weak Interactions (Vol. 52)

Renner, B.: On the Problem of the Sigma Terms in Meson-Baryon Scattering. Comments on Recent Literature (Vol. 61)

Soloviev, L. D.: Symmetries and Current Algebras for Electromagnetic Interactions (Vol. 46)

Stech, B.: Nonleptonic Decays and Mass Differences of Hadrons (Vol. 50)

Stichel, P.: Current Algebra in the Framework of General Quantum Field Theory (Vol. 50)

Stichel, P.: Current Algebra and Renormalizable Field Theories (Vol. 50)

Stichel, P.: Introduction to Current Algebra (Vol. 50)

Verzegnassi, C.: Low Energy Photo and Electroproduction, Multipole Analysis by Current Algebra Commutators (Vol. 59)

Weinstein, M.: Chiral Symmetry. An Approach to the Study of the Strong Interactions (Vol. 60)

Electromagnetic Interactions

Deep Inelastic Lepton Scattering

Drees, J.: Deep Inelastic Electron-Nucleon Scattering (Vol. 60)

Landshoff, P. V.: Duality in Deep Inelastic Electroproduction (Vol. 62)

Llewellyn Smith, C. H.: Parton Models of Inelastic Lepton Scattering (Vol. 62)

Rittenberg, V.: Scaling in Deep Inelastic Scattering with Fixed Final States (Vol. 62)

Rubinstein, H. R.: Duality for Real and Virtual Photons (Vol. 62)

Rühl, W.: Application of Harmonic Analysis to Inelastic Electron-Proton Scattering (Vol. 57)

Experimental Techniques

Panofsky, W. K. H.: Experimental Techniques (Vol. 39)

Strauch, K.: The Use of Bubble Chambers and Spark Chambers at Electron Accelerators (Vol. 39)

Form Factors, Colliding Beam Experiments

Buchanan, C. D., Collard, H., Crannell, C., Frosch, R., Griffy, T. A., Hofstadter, R., Hughes, E. B., Nöldeke, G. K., Oakes, R. J., van Oostrum, K. J., Rand, R. E., Suelzle, L., Yearian, M. R., Clark, B. C., Herman, R., Ravenhall, D. G.: Recent High Energy Electron Investigations at Stanford University (Vol. 39)
Gatto, R.: Theoretical Aspects of Colliding Beam Experiments (Vol. 39)
Gourdin, M.: Vector Mesons in Electromagnetic Interactions (Vol. 55)
Huang, K.: Duality and the Pion Electromagnetic Form Factor (Vol. 62)
Wilson, R.: Review of Nucleon Form Factors (Vol. 39)

Photo- and Electroproduction of Pions

Brinkmann, P.: Polarization of Recoil Nucleons from Single Pion Photoproduction. Experimental Methods and Results (Vol. 61)
Donnachie, A.: Exotic Electromagnetic Currents (Vol. 63)
Drell, S. D.: Special Models and Predictions for Photoproduction above 1 GeV (Vol. 39)
Fischer, H.: Experimental Data on Photoproduction of Pseudoscalar Mesons at Intermediate Energies (Vol. 59)
Foà, L.: Meson Photoproduction on Nuclei (Vol. 59)
Frøyland, J.: High Energy Photoproduction of Pseudoscalar Mesons (Vol. 63)
Furlan, G., Paver, N., Verzegnassi, C.: Low Energy Theorems and Photo- and Electroproduction Near Threshold by Current Algebra (Vol. 62)
von Gehlen, G.: Pion Electroproduction in the Low-Energy Region (Vol. 59)
Heinloth, K.: Experiments on Electroproduction in High Energy Physics (Vol. 65)
Höhler, G.: Special Models and Predictions for Pion Photoproduction (Low Energies) (Vol. 39)
von Holtey, G.: Pion Photoproduction on Nucleons in the First Resonance Region (Vol. 59)
Lüke, D., Söding, P.: Multipole Pion Photoproduction in the s Channel Resonance Region (Vol. 59)
Osborne, L. S.: Photoproduction of Mesons in the GeV Range (Vol. 39)
Pfeil, W., Schwela, D.: Coupling Parameters of Pseudoscalar Meson Photoproduction on Nucleons (Vol. 55)
Renard, F. M.: ρ-ω Mixing (Vol. 63)
Schildknecht, D.: Vector Meson Dominance, Photo- and Electroproduction from Nucleons (Vol. 63)

Schilling, K.: Some Aspects of Vector Meson Photoproduction on Protons (Vol. 63)
Schwela, D.: Pion Photoproduction in the Region of the Δ (1230) Resonance (Vol. 59)
Wolf, G.: Photoproduction of Vector Mesons (Vol. 59)

Quantum Electrodynamics

Olsen, H. A.: Applications of Quantum Electrodynamics (Vol. 44)
Källén, G.: Radiative Corrections in Elementary Particle Physics (Vol. 46)

Field Theory, Light Cone Singularities

Brandt, R. A.: Physics on the Light Cone (Vol. 57)
Dahmen, H. D.: Local Saturation of Commutator Matrix Elements (Vol. 62)
Ferrara, S., Gatto, R., Grillo, A. F.: Conformal Algebra in Space-Time and Operator Product Expansion (Vol. 67)
Jackiw, R.: Canonical Light-Cone Commutators and Their Applications (Vol. 62)
Kundt, W.: Canonical Quantization of Gauge Invariant Field Theories (Vol. 40)
Rühl, W.: Application of Harmonic Analysis to Inelastic Electron-Proton Scattering (Vol. 57)
Symanzik, K.: Small-Distance Behaviour in Field Theory (Vol. 57)
Zimmermann, W.: Problems in Vector Meson Theories (Vol. 50)

Strong Interactions
Mesons and Baryons

Atkinson, D.: Some Consequences of Unitary and Crossing Existence and Asymptotic Theorems (Vol. 57)
Basdevant, J. L.: ππ Theories (Vol. 61)
DeSwart, J. J., Nagels, M. M., Rijken, T. A., Verhoeven, P. A.: Hyperon-Nucleon Interaction (Vol. 60)
Ebel, G., Julius, D., Kramer, G., Martin, B. R., Mühlensiefen, A., Oades, G., Pilkuhn, H., Pišút, J., Roos, M., Schierholz, G., Schmidt, W., Steiner, F., De Swart, J. J.: Compilating of Coupling and Low-Energy Parameters (Vol. 55)
Gustafson, G., Hamilton, J.: The Dynamics of Some πN Resonances (Vol. 61)
Hamilton, J.: New Methods in the Analysis of π-N S Scattering (Vol. 57)
Kramer, G.: Nucleon-Nucleon Interactions below 1 GeV/c (Vol. 55)
Lichtenberg, D. B.: Meson and Baryon Spectroscopy (Vol. 36)
Martin, A. D.: The ΛKN Coupling and Extrapolation below the K̄N Threshold (Vol. 55)

Martin, B. R.: Kaon-Nucleon Interactions below 1 GeV/c (Vol. 55)

Morgan, D., Pišút, J.: Low Energy Pion-Pion Scattering (Vol. 55)

Oades, G. C.: Coulomb Corrections in the Analysis of πN Experimental Scattering Data (Vol. 55)

Pišút, J.: Analytic Extrapolations and the Determination of Pion-Pion Phase Shifts (Vol. 55)

Wanders, G.: Analyticity, Unitary and Crossing-Symmetry Constraints for Pion-Pion Partial Wave Amplitudes (Vol. 57)

Zinn-Justin, J.: Course on Padé Approximants (Vol. 57)

Regge Pole Theory, Dual Models

Ademollo, M.: Current Amplitudes in Dual Resonance Models (Vol. 59)

Chung-I Tan: High Energy Inclusive Processes (Vol. 60)

Collins, P. D. B.: How Important are Regge Cuts? (Vol. 60)

Collins, P. D. B., Gault, F. D.: The Eikonal Model for Regge Cuts in Pion-Nucleon Scattering (Vol. 63)

Collins, P. D. B., Squires, E. J.: Regge Poles in Particle Physics (Vol. 45)

Contogouris, A. P.: Certain Problems of Two-Body Reactions with Spin (Vol. 57)

Contogouris, A. P.: Regge Analysis and Dual Absorptive Model (Vol. 63)

Dietz, K.: Dual Quark Models (Vol. 60)

van Hove, L.: Theory of Strong Interactions of Elementary Particles in the GeV Region (Vol. 39)

Huang, K.: Deep Inelastic Hadronic Scattering in Dual-Resonance Model (Vol. 62)

Landshoff, P. V.: Duality in Deep Inelastic Electroproduction (Vol. 62)

Michael, C.: Regge Residues (Vol. 55)

Oehme, R.: Complex Angular Momentum (Vol. 57)

Oehme, R.: Duality and Regge Theory (Vol. 57)

Oehme, R.: Rising Cross-Sections (Vol. 61)

Rubinstein, H. R.: Duality for Real and Virtual Photons (Vol. 62)

Rubinstein, H. R.: Physical N-Pion Functions (Vol. 57)

Satz, H.: An Introduction to Dual Resonance Models in Multiparticle Physics (Vol. 57)

Schrempp-Otto, B., Schrempp, F.: Are Regge Cuts Still Worthwhile? (Vol. 61)

Squires, E. J.: Regge-Pole Phenomenology (Vol. 57)

Symmetries

Barut, A. O.: Dynamical Groups and their Currents. A Model for Strong Interactions (Vol. 50)

Ekstein, H.: Rigorous Symmetries of Elementary Particles (Vol. 37)

Gourdin, M.: Unitary Symmetry (Vol. 36)

Łopuszański, J. T.: Physical Symmetries in the Framework of Quantum Field Theory (Vol. 52)

Pauli, W.: Continuous Groups in Quantum Mechanics (Vol. 37)

Racah, G.: Group Theory Spectroscopy (Vol. 37)

Rühl, W.: Application of Harmonic Analysis to Inelastic Electron-Proton Scattering (Vol. 57)

Wess, J.: Conformal Invariance and the Energy-Momentum Tensor (Vol. 60)

Wess, J.: Realisations of a Compact, Connected, Semisimple Lie Group (Vol. 50)

Weak Interactions

Barut, A. O.: On the S-Matrix Theory of Weak Interactions (Vol. 53)

Dosch, H. G.: The Decays of the $K_0 - \bar{K}_0$ System (Vol. 52)

Gasiorowicz, S.: A Survey of the Weak Interactions (Vol. 52)

Gatto, R.: Cabibbo Angle and $SU_2 \times SU_2$ Breaking (Vol. 53)

von Gehlen, G.: Weak Interactions at High Energies (Vol. 53)

Kabir, P. K.: Questions Raised by CP-Nonconservation (Vol. 52)

Kummer, W.: Relations for Semileptonic Weak Interactions Involving Photons (Vol. 52)

Müller, V. F.: Semileptonic Decays (Vol. 52)

Pietschmann, H.: Weak Interactions at Small Distances (Vol. 52)

Primakoff, H.: Weak Interactions in Nuclear Physics (Vol. 53)

Renner, B.: Current Algebra and Weak Interactions (Vol. 52)

Riazuddin: Radiative Corrections to Weak Decays Involving Leptons (Vol. 52)

Rothleitner, J.: Radiative Corrections to Weak Interactions (Vol. 52)

Segrè, G.: Unconventional Models of Weak Interactions (Vol. 52)

Stech, B.: Non Leptonic Decays (Vol. 52)

Fluids

Hess, S.: Depolarisierte Rayleigh-Streuung und Strömungsdoppelbrechung in Gasen (Vol. 54)

Langbein, D.: Theory of Van der Waals Attraction (Vol. 72)

Steeb, S.: Evaluation of Atomic Distribution in Liquid Metals and Alloys by Means of X-Ray, Neutron and Electron Diffraction (Vol. 47)

Springer, T.: Quasi-Elastic Scattering of Neutrons for the Investigation of Diffusive Motions in Solids and Liquids (Vol. 64)

Nuclear Physics

Baryon-Baryon-Scattering

Kramer, G.: Nucleon-Nucleon Interactions Below 1 GeV/c (Vol. 55)

DeSwart, J. J., Nagels, M. M., Rijken, T. A., Verhoeven, P. A.: Hyperon-Nucleon Interactions (Vol. 60)

Electron Scattering

Theißen, H.: Spectroscopy of Light Nuclei by Low Energy (70 MeV) Inelastic Electron Scattering (Vol. 65)
Überall, H.: Electron Scattering, Photoexcitation and Nuclear Models (Vol. 49)

Nuclear Moments

Donner, W., Süßmann, G.: Paramagnetische Felder am Kernort (Vol. 37)
Zu Putlitz, G.: Determination of Nuclear Moments with Optical Double Resonance (Vol. 37)
Schmid, D.: Nuclear Magnetic Double Resonance – Principles and Applications in Solid State Physics (Vol. 68)

Nuclear Structure

Arenhövel, H., Weber, H. J.: Nuclear Isobar Configurations (Vol. 65)
Levinger, J. S.: The Two and Three Body Problem (Vol. 71)
Racah, G.: Group Theory and Spectroscopy (Vol. 37)
Singer, P.: Emission of Particles Following Muon Capture in Intermediate and Heavy Nuclei (Vol. 71)
Überall, H.: Study of Nuclear Structure by Muon Capture (Vol. 71)
Wildermuth, K., McClure, W.: Cluster Representations of Nuclei (Vol. 41)

Weak Interactions

Gasiorowicz, S.: A Survey of the Weak Interaction (Vol. 52)
Primakoff, H.: Weak Interactions in Nuclear Physics (Vol. 53)

Optics

Godwin, R. P.: Synchrotron Radiation as a Light Source (Vol. 51)
Hawkes, P. W.: Quadrupole Optics (Vol. 42)
Hess, S.: Depolarisierte Rayleigh-Streuung und Strömungsdoppelbrechung in Gasen (Vol. 54)

Laser

Agarwal, G. S.: Quantum Statistical Theories of Spontaneous Emission and their Relation to Other Approaches (Vol. 70)
Graham, R.: Statistical Theory of Instabilities in Stationary Nonequilibrium Systems with Applications to Lasers and Nonlinear Optics (Vol. 66)
Haake, F.: Statistical Treatment of Open Systems by Generalized Master Equations (Vol. 66)
Schwabl, F., Thirring, W.: Quantum Theory of Laser Radiation (Vol. 36)

Plasma Physics

Geiger, W., Hornberger, H., Schramm, K.-H.: Zustand der Materie unter sehr hohen Drücken und Temperaturen (Vol. 46)
Lehner, G.: Über die Grenzen der Erzeugung sehr hoher Magnetfelder (Vol. 47)
Raether, H.: Solid State Excitations by Electrons (Vol. 38)

Relativity and Astrophysics

Börner, G.: On the Properties of Matter in Neutron Stars (Vol. 69)
Heintzmann, H., Mittelstaedt, P.: Physikalische Gesetze in beschleunigten Bezugssystemen (Vol. 47)
Stewart, J., Walker, M.: Black Holes: the Outside Story (Vol. 69)

Cosmology

Kundt, W.: Recent Progress in Cosmology (Isotropy of 3 deg Background Radiation and Occurence of Space-Time Singularities) (Vol. 47)
Kundt, W.: Survey of Cosmology (Vol. 58)

Solid-State Physics

Schramm, K.-H.: Dynamisches Verhalten von Metallen unter Stoßwellenbelastung (Vol. 53)

Crystals

Behringer, J.: Factor Group Analysis Revisited and Unified (Vol. 68)
Lacmann, R.: Die Gleichgewichtsform von Kristallen und die Keimbildungsarbeit bei der Kristallisation (Vol. 44)
Langbein, D.: Theory of Van der Waals Attraction (Vol. 72)
Ludwig, W.: Recent Developments in Lattice Theory (Vol. 43)

Electrons in Crystals

Bauer, G.: Determination of Electron Temperatures and of Hot-Electron Distribution Functions in Semiconductors (Vol. 74)
Bennemann, K. H.: A New Self-consistent Treatment of Electrons in Crystals (Vol. 38)
Daniels, J., v. Festenberg, C., Raether, H., Zeppenfeld, K.: Optical Constants of Solids by Electron Spectroscopy (Vol. 54)
Fischer, K.: Magnetic Impurities in Metals: the s—d exchange model (Vol. 54)
Raether, H.: Solid State Excitations by Electrons (Vol. 38)
Schnakenberg, J.: Electron-Phonon Interaction and Boltzmann Equation in Narrow Band Semiconductors (Vol. 51)

Magnetism

Fischer, K.: Magnetic Impurities in Metals: the *s—d* exchange model (Vol. 54)

Schmid, D.: Nuclear Magnetic Double Resonance – Principles and Applications in Solid State Physics (Vol. 68)

Stierstadt, K.: Der Magnetische Barkhauseneffekt (Vol. 40)

Optical Properties of Crystals

Bäuerle, D.: Vibrational Spectra of Electron and Hydrogen Centers in Ionic Crystals (Vol. 68)

Borstel, G., Falge, H. J., Otto, A.: Surface and Bulk Phonon-Polarisations Observed by Attenuated Total Reflection (Vol. 74)

Daniels, J., v. Festenberg, C., Raether, H., Zeppenfeld, K.: Optical Constants of Solids by Electron Spectroscopy (Vol. 54)

Excitons in High Density, *Haken, H., Nikitine, S.* (Volume Editors). Contributors: *Bagaev, V. S., Biellmann, J., Bivas, A., Goll, J., Grosmann, M., Grun, J. B., Haken, H., Hanamura, E., Levy, R., Mahr, H., Nikitine, S., Novikov, B. V., Rashba, E. I., Rice, T. M., Rogachev, A. A., Schenzle, A., Shaklee, K. L.* (Vol. 73)

Godwin, R. P.: Synchrotron Radiation as a Light Source (Vol. 51)

Pick, H.: Struktur von Störstellen in Alkalihalogenidkristallen (Vol. 38)

Raether, H.: Solid State Excitations by Electrons (Vol. 38)

Quantum Statistics

Agarwal, G. S.: Quantum Statistical Theories of Spontaneous Emission and their Relation to Other Approaches (Vol. 70)

Graham, R.: Statistical Theory of Instabilities in Stationary Nonequilibrium Systems with Applications to Lasers and Nonlinear Optics (Vol. 66)

Haake, F.: Statistical Treatment of Open Systems by Generalized Master Equations (Vol. 66)

Semiconductors

Bauer, G.: Determination of Electron Temperatures and of Hot-Electron Distribution Functions in Semiconductors (Vol. 74)

Feitknecht, J.: Silicon Carbide as a Semiconductor (Vol. 58)

Grosse, P.: Die Festkörpereigenschaften von Tellur (Vol. 48)

Schnakenberg, J.: Electron-Phonon Interaction and Boltzmann Equation in Narrow Band Semiconductors (Vol. 51)

Superconductivity

Lüders, G., Usadel, K.-D.: The Method of the Correlation Function in Superconductivity Theory (Vol. 56)

X-Ray, Neutron-, Electron-Scattering

Steeb, S.: Evaluation of Atomic Distribution in Liquid Metals and Alloys by Means of X-Ray, Neutron and Electron Diffraction (Vol. 47)

Springer, T.: Quasi-Elastic Scattering of Neutrons for the Investigation of Diffusive Motions in Solids and Liquids (Vol. 64)